A New Vision of the Early Universe

Second Edition

Also by
Robert J. Conover

Journey of the Universe

Second Edition

A New Perspective on its Past,
Present, and Future Evolution

A New Vision of the Early Universe

Second Edition

The Origin of SMBHs, Dark Energy, Fields, Forces, & Particles, and How Each Evolved Naturally

Robert J. Conover

InPerspective Publications

Copyright © 2024 by Robert J. Conover

All Rights reserved. Except as permitted under U.S. copyright law, no part of this publication may be reproduced, stored in a retrieval system, or transmitted in any form, or by any means, electric, mechanical, photocopying, or otherwise without the express written permission of the publisher.

The author and publisher apologize to any copyright holder if permission to publish any copyrighted material has not been obtained. If any copyright material has not been acknowledged, please contact the publisher so that we may rectify in any future reprints.

Cover image from Pixabay.com
Illustrations of photon & neutrino by Anna Keene
Editing by Donna D. Lawson

ISBN-13: 978-0-9797298-1-2 (Hardback)
ISBN-13: 978-0-9797298-6-7 (Paperback)

Library of Congress Control Number: 2023951513

Second Edition
Printed in the United States of America
January 2024

Science/Space Sciences/Cosmology
Science/Physics/Astrophysics
Philosophy/Metaphysics

InPerspective Publications
PO Box 805
San Luis Obispo, CA 93406
www.inperspective.space

To my friends Socrates, Newton, Einstein, and others too numerous to name.

In appreciation for the many hours I have spent with them, listening to their words, analyzing their thoughts, learning their methods, imagining their perspectives, embracing their inspirations, and feeling their encouragement; in pursuit of the truth.

A friendship that has been the core of my happiness for so long, and penetrates by heart so deeply, it is beyond words.

Contents

Introduction	i
One - **The Origin of SMBHs & The Cosmic Web**	1
Initial Observations 1	
The Creation of Matter 8	
Blackholes 11	
Supermassive Blackholes 13	
Alternative Theory for SMBH Creation 17	
Two - **The Origin of Particles & Dark Energy**	29
Early Elementary Particles 30	
The Mystery of the Missing Anti-Matter 36	
The Origin of Dark Energy 40	
Three - **The Origin of Fields**	47
Particle Fields 48	
A Particle's Personal Field 53	
The Double-Slit Experiment 56	
Four - **The Origin of Forces**	63
The Theoretical Basis for Force 64	
The Magnetic Force 69	
The Electric Force 72	
The Gravitational Force 78	
The Strong Force 83	
The Weak Force 86	
Five - **The Origin of Composite Particles**	91
The Cosmic Environment 93	
Cosmic Particle Attributes 101	
The Origin of Particle Mass 105	
Virtues of Composite Elementary Particles 112	

Six - **The Origin of the Atom**   117
    Connector Configurations 118
    Quark Configurations 122
    Meson Configurations 128
    Electron Configuration 129
    Boson Configuration 131
    Photon Configuration 132
    Evolution of a Nucleus 136
    Evolution of the Atom 138
    The CMB 140

Seven - **Speculations on Unsolved Mysteries**   145
    Speculations on the Origin of Dark Matter 145
    Speculations on the Origin of Dynamic Gravity 148
    Speculations on Particle Creation & Decay 151
    Speculations on Why Space & Time are Relative 155
    Mathematics and Uncertainty 165

Eight - **The Origin of Reality**   169
    Finding Reality 170
    Quantum Theory 172
    Uncertainty 174
    Particle Entanglement 182
    The Wavefunction 185
    The Reality of Quantum Theory 186
    A Theory of Everything 191

Nine – **New Perspectives**   195
    The Penergy Perspective 195
    The CAGI Perspective 201
    The Tor Model Perspective 203
    The Mathematical Perspective 208
    The Life Perspective 210
    The Cosmic Perspective 212
    A Cosmic Reality 213

Glossary   217
Notes & References   223
Bibliography   241
Index   247
About the Author   254

Introduction

You are about to embark on an amazing reconstruction project, one that will span the entire universe starting at its birth. Together we will build a new vision of what that epoch looked like. Let's begin by examining why a reconstruction is necessary. What is wrong with our current vision of the early universe? The short answer is that if we theorize the wrong initial conditions and events at its birth, the consequent theories and mathematics supporting those events lead us to conflicts and unanswered questions, which is the position we find ourselves in today. The story of our universe needs a new foundation.

Current theory has our universe beginning as pure energy that expanded from an infinitely small point, creating a hot, dense, matter state that gradually cooled and set the stage for the evolution of our universe as we know it today.[1] That scenario is called the Hot Big Bang Theory.

That theory is a part of the Standard Model of Cosmology. Since it and the Standard Model of Particle Physics each borrow heavily from each other in how they view the early universe, they will henceforth be referred to together simply as the Standard Model. The Hot Big Bang Theory is supported by the following observational evidence.[2]

- Given the observed expansion rate we can project back 13.8 billion years to a time when the universe was much smaller.
- 380,000 years post big bang, atoms formed allowing photons to scatter producing the Cosmic Microwave Background we observe today.
- Hydrogen and helium gases thereafter formed in the ratios predicted.
- The formation of a cosmic web is predicted mathematically and confirmed by computer simulation.

True enough, that observational evidence supports the notion the universe began as a tiny speck of expanding pure energy from which particles, fields, and forces were created and evolved. But that same evidence can support a variety of different beginnings, including the beginning espoused herein.

The evidence describes events that occurred 380,000 years after the big bang. How the universe evolved during the first 380,000 years is beyond scientific inquiry and remains a conjecture. The story of particles, fields, and forces popping into existence within the first second is just that – a story. That conjecture has worked for many years and has inspired an edifice of theoretical support, but it stretches credulity and leaves too many unanswered questions. Complex systems found in nature such as those don't suddenly *pop* into existence, they *evolve*.

Scientists have pointed out that there is a growing collection of observations that simply cannot be squared with today's theories.[3] Physicist Laura Mersini-Houghton, a leading expert on the origin of the universe, tells us that the odds of our universe forming with a big bang at high energies have been calculated to be nearly zero.[5] Astronomer Chris Impey tells us, "...the Standard Model is a bit like an aging movie star whose best work is decades old and whose flaws once seemed slight but are now becoming glaring."[5] Given these observations, it appears that some of the presumptions underpinning our models, though long held and mathematically supported, may be wrong.

The world is ready for a new theory, one that naturally evolves from the initial conditions, rather than from unfounded assumptions. Physicist Harry Cliff tells us, 'This is a moment to reexamine our assumptions and look at old problems from a different angle. More than anything, it is time to put our grand ideas and preconceptions to one side and listen carefully to what nature is saying.'[6]

We have been steeped in the language of the Big Bang Theory for so long it will be difficult to think in other terms. What I am proposing here is that we keep what we know about the universe *after* the first 380,000 years but reconsider what we believe before that time.

Our new vision of the early universe will start with a high-density speck of energy that begins expanding, but without any preconceived notions as to what follows. We will build an entirely new early universe from that meager state without adding any new dimensions, particles, forces, or anything else that is not already known to science; a beginning that starts differently but naturally *evolves* into exactly the universe we see today.

For some readers, exploring a new theory without a mathematical basis will seem like a waste of time. Those readers are reminded that some of the greatest accomplishments in physics, such as Newton's theory of gravity, Maxwell's theory of electromagnetism, and Einstein's theory of relativity, all started from a theoretical basis before the complete mathematical structure was finalized. As Physicist David Lindley says in relating the story of Faraday's discovery of the *field*, "It's important to understand that conceptualization came first, the mathematics later. Mathematics is the language in which physical ideas are conveyed but not the origin of those ideas."7

Much of the Standard Model's mathematics will easily translate to the model we create. For the most part, the new model does not take issue with the Standard Model's mathematics, only the *interpretation* of what that mathematics means. Consequently, this story can be told without referring to a new mathematical structure.

Physicist Lisa Randall informs us '...ultimately critical thinking is the only reliable way to answer questions about the makeup of the universe.'8 We will employ that critical thinking as much as possible, recording our observations and making reasonable deductions. Our first two deductions in Chapter One are noted as *Tentative Deductions* because, though quite reasonable, their support is minimal, and they represent a notable departure from current theory. Our next three deductions, however, are well supported with a significant amount of observational evidence that validates the first two deductions.

The first six chapters cover the Origin of: 1. Supermassive Blackholes & the Cosmic Web; 2. Particles & Dark Energy; 3. Fields; 4. Forces; 5. Composite Particles; and 6. The Atom. The origin of each of these phases *naturally follows* from the phase before it because that is the order in which the universe *evolved*. That is a very important point, so I am going to repeat it. Each of these phases naturally follows from the phase before it because that is the order in which the universe evolved.

We shall undertake this reconstruction project embracing the following four rules.

1. **No extraneous initial conditions**. Having observed mass equals energy and the universe is expanding, scientists have concluded that *the universe began as a speck of near-infinite-density pure energy that began expanding*. We need no other initial conditions.

2. **No magic**. Nothing just suddenly pops into existence. All creations have a cause, all causes have a foundation, and all phases evolve from the phase before it.

3. **No Math**. Math is a wonderful tool, but when developing a new theory, it can lead us astray. As observed, Newton, Maxwell, and Einstein all developed complex ideas about the universe before completing the math to support them. Astrophysicist David Lindley described Galileo as a thinker who prized empirical information as the starting point for any true scientific investigation, and he tells us that Galileo's style of investigation "insists that mathematics must be the servant, not the master."[9] We shall utilize that style as well.

4. **Keep an open mind.** Anything is possible, as long as it is supported by observational evidence and/or reasonable deductions.

Armed with only an open mind and a blank slate, we shall endeavor to make good observations and deductions. We shall also stay in keeping with the scientific goals of creating a model that is simple, clear, natural, and elegant.

This book, like the universe, starts with simplicity and evolves to complexity. Though readers with a Ph.D. will glean more from the text, it is written for comprehension by the general reader with an interest in the evolution of the universe. I'm confident it will be a treasure to you both.

We now boldly go where no one has gone before, to the dawn of the universe armed with only a blank slate and no expectation as to what we will find there. Our journey begins!

INTRODUCTION

Chapter One
The Origin of Supermassive Blackholes & The Cosmic Web

Initial Observations

Our mission is to build a new vision of our early universe using only our keen observations and deductions, limited by the four rules expressed in the Introduction. We start with only two initial conditions: 1) *the universe began as a speck of pure energy,* and 2) *it began expanding.* Let's begin by observing the gross aspects of the universe we see all around us. We observe galaxies, blackholes, stars, planets, moons, and gases, all of which we simply characterize as matter. But what is matter, really?

Over the past century, scientists have been drilling into matter exposing its layers. An early examination confirmed the long-held belief that it was comprised of tiny bits called atoms. Drilling deeper scientists discovered atoms are comprised of a nucleus orbited by different configurations of electrons. Drilling into the nucleus revealed it to be made up of two particles labeled Protons and Neutrons. Going deeper they found that both are in turn made up of two different particles called Quarks. We do not yet have the technology to drill deeper into matter beyond the quark, but many physicists believe deeper layers are possible. The known particles within the atom are referred to as *subatomic particles*, hereinafter called *Sub-A's*.

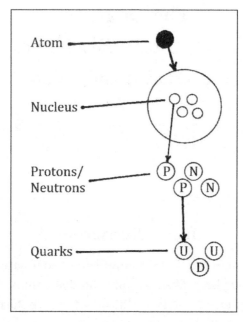

Figure 1.1 The drilling into matter from the atom down to quarks.

Matter is important to us not only because we are made of it, but because it tells us something very important about the universe. According to Einstein's famous equation $E=MC^2$, energy (E) is equal to mass (M) times the speed of light (C) squared. Mass is the measurable content of matter. It means energy and matter are equivalent; everything that is matter is made from this stuff called energy. We know what matter looks like, but what is energy?

Energy

Energy is known in two forms, physical energy, which is related to mass, and *work energy*, which is related to particle movements and fields. For scientists, the term energy generally refers to one of the several manifestations of work energy, such as radiant energy or nuclear energy. Presently, I am referring to physical energy from which everything with mass is made; often called pure energy. For clarity, and to separate it from other forms of energy we may discuss later, pure energy will be called *Penergy*. So, Penergy is the energy that Einstein referred to in his famous equation $E=MC^2$, all of which came from the initial speck of pure energy and is the energy from which everything we observe in the universe is made.[1]

Penergy is otherwise difficult to define because there is nothing else like it. It's not a solid, liquid, or gas, but it has the capacity to make itself into things we can refer to as such. It is very dynamic, meaning it has the capacity to *be* many things (*blackholes and particles*); to make things *happen* (*evolution and forces*); to *bring about* non-tangible things (*consciousness and intelligence*); and to possess even more capabilities we are only now beginning to realize. The reader will gain a better appreciation for what Penergy is as its many dynamics and characteristics are examined in due course.

Penergy can appear in several *states* or conditions, which will be discussed shortly. Initially it is referred to as being a cloud, but that is only to give the reader a visual image. For the moment, simply think of Penergy as a unique, fluid-like substance, as it is imagined in most cosmological models.[2] It is the only real substance in our entire universe and appears in many manifestations. We will make an in-depth examination of Penergy and how it relates to other forms of energy in Chapter Nine.

We have observed that the entire visible universe is made of matter, and that matter is made of Penergy. To better understand the landscape of our universe we might next ask, *where did Penergy come from?*

Penergy

The short answer is we don't know where the Penergy came from. That question leads to what existed before the big bang, which thus far is beyond scientific inquiry. We can however make reasonable speculations on the Penergy's initial size and density.

We have observed evidence indicating galaxies in our universe are moving away from each other. We can infer the universe is expanding, not just at its edges like a balloon, but from within, like a loaf of raisin bread in the oven expands with all its raisins moving away from each other. In this sense the galaxies aren't really moving on their own. Their only movement is incidental to the space between them expanding, which occurs in all three dimensions, so we observe this as galaxies seemingly moving away from each other.

If the entire universe is expanding, it means that sometime in the past it was much smaller. Scientists have calculated the universe to be 13.8 billion years old and according to Physicist Ben Still, "It is thought that our Universe began its life as pure energy which expanded from an infinitely small point."[3]

Astronomer Philip Ball portrays the beginning as the entire universe being smaller than an atom.[4] Astrophysicist Neil deGrasse Tyson tells us that at the beginning of time, all space, matter, and energy in the universe fit within a pinhead.[5] Most authors on the subject say about the same thing, so let's just agree the initial speck of Penergy was very small. Some say that it was *infinitely* small. But what does that mean?

The Singularity Problem

Within weeks of Einstein publishing his theory of General Relativity in 1915, Physicist Karl Schwarzschild used Einstein's equations to demonstrate that an object in space with sufficiently strong gravity could create something we now call a blackhole. Soon thereafter, Russian Mathematician Alexander Friedmann found solutions to Einstein's equations describing universes evolving from a singular state of infinite density – what is now called a *singularity*.

In the 1970's, Physicist Stephen Hawking and Mathematician Roger Penrose wrote a theorem concluding that the energy kernel at the start of an expanding universe such as ours, and at the core of all blackholes, is a singularity. Hawking later wrote that he changed his mind and that there was no singularity at the start of the universe.[6] Still, the idea of the universe starting from a singularity with infinite energy and density persists. Astrophysicist Carolyn Devereux tells us in her recent book, *Cosmological Clues,* that cosmologists do not like the singularity idea but do not know how else to describe the start of the universe and they put up with it until finding something better.[7]

The word *infinite* here has its usual meaning – something without limit that goes on forever. We can imagine something infinite, like a number line to which one can always add one more number, but all such examples are simply human-made concepts, not reality. Infinity is an interesting idea and may even be useful in mathematical calculations, but to believe that anything in nature goes on forever is a challenge. We have not observed a singularity or anything else in nature that is infinite. Infinite density and anything else one might deem infinite is only theoretical.

Infinities create their own challenges for scientists. As Astronomer Chris Impey tells us mathematics has many ways to manipulate and deal with infinities, but in physics infinity is a big problem.[8] Physicist Michio Kaku tells us that to a physicist, infinity is just a sign that the equations are not working; that the physicists don't understand what is happening.[9] Einstein believed the presence of singularities was a sign of imperfect physical understanding and that it made no sense for an object to have zero size and infinite mass density.[10]

The concept of singularity seems to present a conflict in ideas. If something has *infinite* density, how is it ever going to change short of a God-like intervention. If its density is infinite, it sounds pretty hard, unbreakable, and unchangeable. If its density is infinite, it can't form anything else. If it becomes something else, its density must not have been infinite.

Anything infinite is an abstraction, purely a mathematical object. As we know, mathematics can be wonderful, but its equations don't always represent reality. As Michio Kaku again tells us, we cannot blindly accept the math of Einstein's theory since his equations predict the center of a blackhole or the beginning of time as infinite, which makes no sense.[11] For all these reasons we will keep things simple and not infuse any infinities or singularities into our project here. Our blueprint for constructing our new vision of the early universe will call for a non-singularity; a simple *speck* of Penergy will do.

We don't need to resolve how dense the speck of Penergy was but agree it was very dense; a condition I like to call the NIB state - **N**ear **I**nfinite **B**ut (not quite). For reasons unknown, the speck of very-dense Penergy changed and the point in time that change occurred we call the big bang. We don't know how the speck of Penergy originated, how long it existed prior to becoming our universe, or what caused it to change. Any speculations on those questions would only detour us down a philosophical path away from our goal so we won't go there.

We know little about the speck of Penergy beyond the fact that apparently our entire observable universe is made from it. Anything else we learn about it will have to be derived from our observations – measured, deduced, or surmised. The next observation we might make is that the Penergy somehow turned itself into matter. That observation of course brings up the questions of *when, how,* and *from what density?* Let's see what observations we can make that may answer these important questions.

The Penergy Density Question

The Standard Model posits that in the first millionth of a second of the big bang all the basic elements and forces were formed, creating the Sub-A's we know today.[12] In all due respect for this Hot Big Bang Theory, the story of all those things simply popping into existence is difficult to believe. In fact, further observations about Penergy that will be related throughout these first chapters suggest the universe started quite differently.

Matter simply popping into existence makes no sense. If Penergy started out as a high-density speck and somehow changed to a mass of comparatively light-density particles, it is more reasonable that it would have accomplished that feat over time in a series of steps, rather than in a single, sudden, explosive episode. Going from one extreme of a near-infinite density speck, to another extreme of various, complex, light-density particles, inside an explosion lasting a fraction of a second *might* be possible, but is unsupported, unreasonable, and unnecessary. This is a core concept for a new vision of the early universe and a significant departure from current theory, so let's examine it closely.

Let's begin by recognizing that the primordial universe at the big bang is a mystery beyond the reach of science's physical models.[13] Its story is purely theoretical and can only be told by piecing together good observations and deductions. As noted in the introduction, all the observable evidence supporting the big bang theory occurs 380,000 years after the big bang, and none of that evidence supports the notion of matter, fields, and forces suddenly popping into existence.

Let's imagine a new big bang without anything popping into existence. The moment the speck of pure energy (*Penergy*) began expanding it would have instantly begun losing density. The fact that mass and energy as we know it today can be easily converted into each other suggests that the pure energy density implied in $E=MC^2$ is much lighter than a near infinite density. The apparent change from a near-infinite density to a relatively light particle-density suggests that Penergy is capable of existing in a range of densities.

Based on the phase transitions (*solid→liquid→gas*) we know matter to be capable of, it is reasonable to believe that multiple, specific, and dynamic density phases could take place both at the maturing of a blackhole and at the unfolding of a universe. Why would something so important and dynamic as Penergy, the substance of our entire universe, be limited to existing in only two density states? And two densities that are not close to or even related to each other? Penergy having a density makes sense, but Penergy being confined to only two density-states makes no sense. Given these arguments (*and more to follow*), we can make our first tentative deduction.

> Tentative Deduction #1: If Penergy initially existed in a near infinite-density state as believed, and we know it to exist in a comparatively light-density state as particles, we can infer that Penergy can exist in a range of densities. Additional support for this deduction will be revealed in subsequent observations and deductions.

The Creation of Matter

The standard Model's story of complex forces and particles popping into existence within the first second of the big bang stretches credulity. If all matter that exists today was created within the first minute of the big bang, the universe would have been very tiny, and it is difficult to imagine how all of that matter could have been created and compressed down to such a small space. Astrophysicist John Gribbin tells us that, "...everything we can see would have been crowded into a volume of space just one millimeter across. It sounds ridiculous, but that is what the observations of the expanding universe and the equations of general theory of relativity tell us."[14] That may somehow work mathematically, but as Dr. Kaku expressed earlier, we cannot blindly accept the equations of Einstein's theory that makes no sense. It does sound ridiculous, and if you are open to starting the universe from a non-singularity, it also seems unnecessary.

Protons and neutrons are complex, believed to be made up of two different types of quarks and a sea of eight kinds of gluons. Why would these complex particles, along with electrons, neutrinos, and photons, all be created inside the first minutes of the big bang with no purpose, cause, blueprint, or evolutionary basis?

It makes much more sense for particles, especially complex-composite particles such as protons and neutrons comprised of many different quarks and gluons, to have an evolutionary origin. Again, complex systems found in nature such as those don't suddenly *pop* into existence, they *evolve*. How that could have happened will be addressed in the following chapters. For now, the point is that a near-infinite density state of Penergy suddenly changing into a comparatively light-density state of complex particles is unsupported and unreasonable, especially since there are more rational alternatives. This leads us to our second tentative deduction.

> Tentative Deduction #2: If Penergy initially existed in a near infinite-density state, there is no compelling reason to believe it suddenly changed into a mass of many different types of light-density, complex particles within the first minutes of the big bang, as theorized. In short, our subatomic particles were not likely created at the big bang. Additional support for this deduction will be revealed in subsequent observations and deductions.

If the near-infinite speck of Penergy did not convert to particles instantly at the big bang, it suggests the Penergy had to await a lesser density for light-density particles to form. For the Penergy density to lessen, the volume of the universe would have to grow. Our baby universe of Penergy would have to expand, as we know it does.

Penergy Density Scale

The notion of density derives from the relationship between substance and volume. Density is the degree of compactness of a substance in a given volume. Think of high density as lots of stuff packed into a small volume.

As the universe expanded in volume, the Penergy substance would have become thinner in density. With regards to Penergy however, it is believed that the thinning was not in direct proportion to its expansion. Scientists tell us that the energy expanding the universe is *growing* as it expands.[15] That means that as it expands from within, (*like raisin bread*) more Penergy is produced, thus it's not like a fixed amount growing thinner with more volume. With Penergy, the thinning density rate is much slower than its expansion rate. As the Penergy thins, more Penergy is pulled out of the existing Penergy, and apparently the thinner it gets the more pressure there is to extract more.

This seems to be a violation of the first law of thermodynamics, which states that energy cannot be created or destroyed. In this case it is not that new energy is being created out of nothing. It is simply the dynamics of how Penergy expands disproportionate to its losing density by drawing more out of itself. It's not the expanding universe that is causing the thinning, it is apparently the thinning of the Penergy pushing out to expand and create whatever volume it needs. This is largely theoretical and for our project we need not pursue it further.

Since we currently have the universe expanding and the Penergy density thinning, it would help us to get a handle on some relative measurement of the density. We can make up a density scale to track the phase transitions the Penergy may have gone through. If the near-infinite NIB-state density was given a value of 1.0, we could break the scale down into increments of .1, giving it .9, .8, .7 densities, etc.

If the 1.0 represents a near-infinite density (*near solid state*) of Penergy, then the .9 and .8 densities that came into existence as it thinned would still be pretty dense, perhaps too dense for anything significant to have happened. At .7 say, the Penergy might be thin enough to allow things to happen with which we can identify. During the thinning from 1.0 to .7 density the universe could have grown substantially in size, especially if the expansion rate was in some way proportional to the energy density as some physicists have suggested.[16]

Some readers may think the Tentative Deductions we've made so far are borderline and wonder if there is additional observational evidence to support those deductions. Yes, there is. It is strong evidence, and it comes from some of the oldest creations in the universe – supermassive blackholes. Their existence, as well as other observations and deductions we will make later, all lend a great deal of validity to our initial Tentative Deductions.

Blackholes

Most of the universe is empty space except for various concentrations of huge galaxies, each made up of a supermassive blackhole surrounded by billions of stars. What can we learn from this observation? First let's look at the elephant in the picture – the blackhole.

A blackhole is a region of space where gravity is so strong that nothing can escape its boundary called the *event horizon*. Blackholes were predicted from Einstein's field equations, for many years believed to possibly exist, and finally in the 1990's they were confirmed to exist.

The blackhole creation theory is that a mass with sufficient compaction (*density*) could have enough gravitational influence to turn it into a blackhole. To get an idea of how much compaction we are talking about, for the planet earth to become a blackhole it would have to have its entire mass compressed down to a sphere the size a little smaller than a ping-pong ball.[17]

A star several times bigger than our sun when it reaches the end of its life is theorized to implode into a much smaller, denser mass that goes through various stages, condensing further and further, eventually becoming a blackhole. It is theorized that this stellar collapse into a blackhole is the normal life cycle of stars that are a few times to one hundred times bigger than our sun. There are relatively few that size however, less than one per cent of the observable stars.

Blackholes make it apparent that Penergy has the capacity to both expand and condense. It naturally expands when in its thinning, static-state, and condenses when a glob of it is in a very dense state and has the opportunity to *spin*, giving it a gravitational influence.

If the initial Penergy speck started out in a state of very high density, we might imagine it as a very fast spinning puck with a very strong gravitational field. For reasons unknown, the puck apparently broke apart or ruptured and spewed its pure energy. It naturally would have attempted to regain its spin rate and strong gravitational influence, but if the rupture was too great to overcome, it would have continued spewing and spreading. While the interior portions of the Penergy cloud would have been working to condense, the much thinner-density outer edges of the cloud would have been working to expand. Obviously, the expansion dynamic won the competition, and our nascent universe was born. That expansion continues today.

Blackholes are not yet well understood. According to general relativity, every blackhole contains a singularity, yet inside the blackhole Einstein's field equations become nonsensical. Hanging onto the concept of a singularity might be inhibiting our understanding. Perhaps the inside could be better understood if the calculations were based on its NIB-state rather than a singularity, and its mass based on the density of its Penergy rather than on its apparent gravitational influence. We will examine new concepts of mass and gravity in due course.

Stephen Hawking told us, "Thus the matter inside a blackhole would be trapped and would collapse to some unknown state of very high density."[18] Matter crushed to a very high density would turn it back into Penergy, which would constitute the bulk of the blackhole's content. Accordingly, we might theorize that the particle mass inside a blackhole is being condensed back to Penergy, and the Penergy itself is being condensed further and further until it is presumably back to its NIB-state. Consequently, a blackhole does not actually need a mass of particles to come into existence or sustain itself. According to Astrophysicist Luciano Rezzolla, the Schwarzschild resolution to Einstein's field equations does not require the presence of matter.[19] All of this suggests the mass of a blackhole is more about condensed Penergy than about particle matter.

Supermassive Blackholes

At the heart of all galaxies is a supermassive blackhole (*SMBH*). SMBHs are millions, even billions, times bigger than our sun. Scientists have not yet agreed on a sound theory for SMBH creation. Under current theory they are not supposed to exist. Scientists are hamstrung trying to develop a theory for SMBH creation *after* atoms and stars were created in keeping with the Standard Model's version of the Big Bang Theory.

Trying to base a SMBH creation theory on the Hot Big Bang platform doesn't seem to work. Blackhole growth simply by accretion (*eating surrounding stars and material*) is slow. It would take far too long to grow to supermassive size. Computer simulations have indicated possible complex processes involving a mixture of early gas formations, temperature differentials, density fluctuations, and perturbations, all of which must be just right to form large galaxies. Again, such a complex scenario might be capable of producing a large galaxy, but it is not likely to have created billions of galaxies with a SMBH at their center.

Under the Hot Big Bang Theory, the most feasible way SMBHs could have grown to millions of solar masses is through blackhole mergers. Let's examine that possibility.

Arguments For and Against SMBH Creation by Mergers

The Hubble and James Webb Space Telescopes have produced deep-space images showing galaxies that have been calculated to have formed well within the first billion years after the big bang. This does not allow enough time for stars to form, die, blackholes to form, and for the blackholes to have grown to SMBH size simply from isolated mergers.[20] The merger theory might work, however, if the merging blackholes were uncharacteristically large.

Astronomer Martin Rees developed a theory that larger blackholes could have started from seed blackholes created by the collapse of thousands of stars, and then those larger blackholes could have merged to create SMBHs.[21] This is a rather complex formula for the creation of a very large blackhole that may require a seed mass of as much as 10,000 solar masses.[22] Scientists are now speculating on seed masses of 100,000 solar masses produced from the collapse of early giant gas clouds. Those complex scenarios may have happened somewhere in the universe, but it doesn't seem plausible for the creation of every SMBH at the heart of billions of galaxies.

Computer simulations have been tweaked to the point of allowing very large blackholes to have formed in the early universe. But all those simulations eventually require mergers to take place for SMBHs to be created within a billion years after the big bang. Mergers at that age of the universe would require whole galaxies to have merged over and over to grow to a SMBH size. Again, that is an unlikely prospect for creating a SMBH for the billions of galaxies in the universe.

Observational evidence regarding spiral galaxies with SMBHs suggests their formation could not have happened through a frenzy of galaxy mergers. A spiral galaxy has a center mass with arms growing out around it, as shown in Figure 1.2a.

Figure 1.2a Spiral Galaxy

It is believed that when spiral galaxies collide and their blackholes merge, the process tends to scatter the spiral-disk shape.[23] With multiple mergers, the consolidation would create a diffused elliptical galaxy, as shown in Figure 1.2b.

Figure 1.2b Elliptical Galaxy

Seventy percent of the galaxies in our universe are spiral in shape.[24] The fact that we can observe so many pristine, spiral galaxies containing a supermassive blackhole suggests those galaxies have not suffered the turmoil of multiple mergers and apparently acquired their supermassive blackhole by other means. The remaining thirty percent of galaxies are elliptical and generally house a much bigger supermassive blackhole *billions* of times bigger than our sun. They look like a near spherical cloud, and due to their size and appearance they indeed seem to be the remnants of galaxy mergers. The Milky Way is expected to become an elliptical galaxy once it collides with the Andromeda galaxy.[25]

If SMBHs were not formed by either accretion or mergers, how were they created? Given our Tentative Deduction#1 that the Penergy starting the universe constantly thinned as the universe expanded, there is a very plausible explanation for how these supermassive blackholes came about. The short answer is that they were formed out of Penergy when it was quite dense in the early universe. Let's examine how that may have happened.

Alternative Theory for SMBH Creation

One thing we have observed about blackholes is that they spin, which seems to be a natural dynamic of Penergy. There is plenty of evidence for forms of Penergy naturally spinning. We know that particles spin, planets spin, stars spin, galaxies spin, blackholes spin, and in the case of stars, they start the process naturally on their own accord. Everything in the universe spins or is made of something spinning. It is what Penergy naturally does. With this notion in mind, how would that affect SMBH creation?

If stars can start the process of their own creation by swirling the substance from which they are made, there is no reason to doubt that Penergy can do the same. As the universe expanded and Penergy thinned, at perhaps .7 density the Penergy could have begun breaking up into huge swirls. The developing swirls would be large and quite heavy so the movement would initially be rather slow. But eventually, just like star formation, as the circular motion gradually sped up it would gain a growing gravitational influence and cause the Penergy to pull in toward the center of the swirls, leaving thinner-density areas at their outer edges.

At those outer edges, now at say a .6 density, many more swirls would have begun, drawing in more Penergy, leaving their outer edges less dense. Those outer edges, now at say .5 density, would have created even more swirls, again with their outer edges thinning, setting off a *cascade of swirls* of lesser and lesser densities, as illustrated in Figure 1.3.

Soon there would have been thousands, millions, and billions of swirls of different sizes, causing all kinds of anomalies in their distribution. As the swirls became better defined, they would have begun to create a stronger but still weak gravitational influence that would have slowed the Penergy expansion near them in proportion to their density, causing even more anomalies in their distribution.

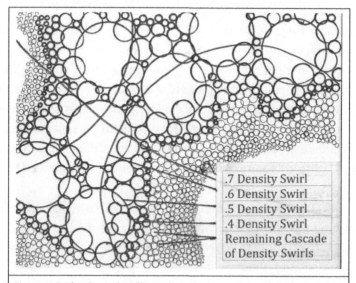

Figure 1.3 The Cascade Effect, showing how each swirl (*depicted by circles*) drew in Penergy, causing its outer edge to thin in density, allowing new, smaller swirls to begin. This effect happened over and over, cascading through smaller and smaller swirls with diminishing densities.

Over time, the bigger swirls would continue condensing, gaining angular momentum with a growing gravitational influence. Eventually the swirls with sufficient density and angular momentum to sustain themselves would become *pure energy blackholes*. These blackholes, many thousand and million times the size of our sun, will eventually be at the core of billions of galaxies making up our universe. The fluctuations, contractions, territorial battles for Penergy, and the *cascade effect*, would cause the resulting galaxies to form in clusters, groups, walls, filaments, and other peculiar distributions that would ultimately shape our universe.

Pure energy blackholes are not yet recognized but this scenario seems to be a very credible resolution to the mystery of their formation. An article in *Quanta Magazine* in December 2017, entitled *Earliest Blackhole Gives Rare Glimpse of Ancient Universe,* discussed the discovery of a very old blackhole with its age placed at 690 million years after the big bang and its size equal to 780 million times the mass of our sun. The researchers' calculations discovered that if the blackhole started out as a collapsed star like a blackhole is normally formed, it would not have had time to grow to its enormous size. They further calculated that if it came into existence soon after the big bang, it would have had to start out about 1000 times the mass of our sun.

The researchers speculated that perhaps hydrogen/helium gases in the early universe condensed into a blackhole, though admittedly this is inconsistent with scientists' current view of the early universe. Since then, scientists have found galactic blackholes measuring several billion times the size of our sun. The JWST is also finding many more large galaxies than expected, which is quite consistent with the cascade theory. Given these reports and the deductions we've made so far, the creation of early, pure energy blackholes seems very plausible.

We now have a universe full of emerging SMBHs made of medium-high density Penergy. At the very end of the cascade of swirls the Penergy density would have thinned to a point where early elementary particles (EEPs) could spin into existence. We will look at them in the next chapter.

The Cascade Effect would have created many swirls of various sizes that spun into existence, but we only see very large supermassive blackholes and very tiny, tightly wound particles. What happened to all the swirls sized in between the two?

SMBH/Particle Relationship with Penergy Density

The larger, medium to high-density swirls would have gained angular momentum and attained sufficient gravitational influence that they would have become SMBHs. As the universe expanded, the Penergy density between swirls would have grown too thin to support the intermediate size swirls having no particle mass and too little gravitational influence to become a stable blackhole. Being unable to support their bulk in the surrounding, thinning density, they would have eventually disintegrated back into the Penergy medium. That disintegration would have appeared much like we know a failed star to slowly fade until it is completely invisible. The disappearing mid-sized swirls would have left huge voids.

At the other density extreme, the tiny, light density swirls became very-tightly wound entities we are calling Early Elementary Particles (*EEPs*). Since they are still with us, they too were obviously capable of sustaining themselves. The first EEPs would have begun combining into larger, first-generation composite particles. These early generations of composite particles would have also disintegrated once the surrounding Penergy grew too thin to support their bulk. There were at least two generations of elementary composite particles that could not survive in the thinning Penergy density, but possibly there were more.

These transitions would eventually leave the universe with only two stable elements: various sizes of SMBHs and tightly wound EEPs. Both elements still exist today in the remaining very-light density Penergy-medium permeating our universe. Apparently, there is an ideal spin-rate to mass/energy ratio that allows for stable elementary creations such as blackholes and particles.

Deduction #3: We have observed that when particles larger than our Sub-A's are created in particle colliders they immediately decay into smaller particles. This implies that those larger particles cannot survive in the very-thin Penergy density level currently pervading our universe. It also explains why all intermediate sized swirls created by the Cascade Effect disappeared leaving large voids.

Evidence of Cascade Effect & SMBH Creation

The hypothesis here is that the distribution of the pure-energy supermassive blackholes at the center of all galaxies was structured by the *cascade effect* described above, making SMBH creation well underway when the elementary particles and stars evolved. Let's examine the observational evidence supporting this hypothesis.

The Number and Size of Blackholes: SMBHs are known to exist in many sizes, having been measured in hundreds of thousands, millions, and billions of solar masses. The vast number of sizes of SMBHs, all far too big to have been created from solar collapse, suggests that the theory of a cascade of blackholes in the early universe is consistent with observations. The disappearing intermediate size swirls at the end of the cascade is the reason we see very large voids and is the reason we do not see blackholes between fifty thousand and one hundred solar masses.

The Cascade Effect should have left some very large galaxies created very early. That is exactly what the JWST is finding as reported in numerous articles. Scientists are stumped to explain the size of galaxies having have come into existence less than 500 million years after the big bang but viewing the early universe from the perspective of Penergy density and the Cascade Effect, they make perfect sense.

The JWST is making new discoveries weekly regarding the age, size, and distribution of the earliest galaxies. In an article published in *Live Science*, Brandon Specker reported that the JWST recently discovered a chain of massive galaxies measuring in the billions of solar masses having been formed about 12 billion years ago. The group has been dubbed "The Vine", and is depicted as a large, bow shaped array that may be part of a larger cluster [*perhaps a circle*]. This and all other discoveries made by the JWST are quite consistent with the expected results of the cascade effect.

Galactic Distribution: Galaxies are not randomly scattered across the universe but are preferentially found in super clusters, some containing thousands of galaxies. Within those super clusters are smaller clusters, different sized groups, walls, sheets, and filaments, with very few isolated galaxies. Within larger individual galaxies are smaller galaxies of various sizes. According to Astronomers Luke Barnes and Geraint Lewis in their book, *The Cosmic Revolutionary's Handbook*, many *dwarf galaxies* have been found in a plane rotating around our Milky Way galaxy. Dwarf galaxies have also been found in a plane rotating around the nearby Andromeda galaxy.[26]

These observations and discoveries support the idea of the Cascade Effect. The cascading swirls left galaxies embedded in clusters and groups down to the level of dwarf galaxies appearing inside larger individual galaxies. The Cascade Effect should have left the smaller swirls more prevalent. This bares out. It is reported that dwarf galaxies are the most common galaxy type in the universe. Astronomer Chris Impey describes the galactic terrain as "littered with dwarf galaxies".[27]

The Cascade Effect should have also left some of the smallest galaxies appearing within the area of the voids. This is exactly what is observed. Astrophysicist Paul Sutter reports that if you zoom in on a portion of the cosmic web and look at the voids you will see dim dwarf galaxies. He goes on to say that upon a close examination of those voids, one sees a faint echo of the cosmic web; "...the cosmic web can be seen as a series of *nested* cosmic webs, each 'level' dimmer and smaller than the last."[28] That description sounds very much like observational evidence for the Cascade Effect.

Galactic Movement: Paul Sutter also tells us that the galaxies comprising the cosmic web are moving. Their pattern is not static. "Galaxies are buzzing around in orbits with the cluster."[29] This observation is consistent with the Cascade Effect. Many of the galaxies were created while their Penergy was in a swirling motion being dragged around by a larger swirl. There is no reason to believe those SMBHs that eventually formed would have ceased that circular motion, which could easily be interpreted today as *galaxies buzzing around in orbits.*

Galactic Halos: Astrophysicist John Gribbin reports astronomical observations of *deem blue galaxies,* which are described as very numerous and very faint dwarf galaxies that existed billions of years ago but have since disappeared as evidenced by there being no younger galaxies of that type. Those faint dwarf galaxies are so numerous they have been described as 'cosmic wallpaper'.[30] Dr. Gribbin's description describes the swirls just below the threshold for SMBH stability. They would have existed long enough for particles to come into existence but ultimately as the Penergy density thinned those early smaller galaxies could not sustain themselves and would have suffered a force decay (*discussed shortly*) back into the Penergy medium, leaving faint evidence of their existence.

Galactic Shape: It is observed that both galaxies and solar systems are generally found in the shape of a disk. This may be because they both form in a very similar manner.

Stars form from a high concentration of gas and dust collecting, rotating, and drawing in material until the ball of gas flattens into a disk. Once the star's nuclear fusion begins, the band of residual matter at the outer edge of the star's disk coalesces into planets circling in the same plane creating a solar system.

Similarly, by the time the cascade drew down to the smallest swirls, the large swirls had picked up sufficient angular momentum to cause the adjacent Penergy and matter to flatten into a disk, ultimately creating the vast number of spiral galaxies we see today. As newly created particles combined into atoms and gas, the atoms and gas swirled into stars. Once the now SMBH gained sufficient angular momentum it would have pulled in the surrounding gas, matter, and stars, encircling them in the same plane as the original disk.

In summary, just as planets were formed in a plane incidental to star formation, stars were formed in a plane incidental to SMBH formation. This also implies that the SMBH had started forming well before the atoms and stars were created.

Relationship between Galactic Stars and SMBH: Astronomer Chris Impey tells us that every galaxy has a SMBH at its center and the mass of the blackhole is tightly coupled to the mass of the stars in that galaxy.[31] This observation suggests the blackhole was already developing when particles and stars evolved, and once established, the blackhole drew in just the correct amount of matter that it could support with its gravitational field. He also points out the mystery surrounding the tight relationship between the mass of a SMBH and the mass of all the old stars in its gravity. He says, it's as if the blackhole "knows what kind of galaxy it lives in."[32] This mystery is solved if you imagine SMBHs and particles developing near simultaneously and once established the blackhole draws in just the correct amount of matter its gravitational field could support.

Matching our Cosmic Structure – the Cosmic Web: As observed, galaxies are not spread uniformly across the universe but are gathered in clusters, groups, and filaments. The overall arrangement has a honeycomb appearance and is commonly referred to as the Cosmic Web. It is described by Wallace H. Tucker in his book, *Chandra's Cosmos,* as "massive galaxy clusters as nodes that are interconnected through an intricate web of filaments and sheets of tenuous gas and galaxies, with nearly empty regions called voids taking up most of the volume."[33]

That description is depicted in Figure 1.4 below. A reconstruction of the cosmic web formation has been computer simulated resulting in a complex theory of energy movements and perturbations, having little mention of the contribution of blackholes. Let's examine an alternative theory that accounts for the creation of the cosmic web in a more natural and less complicated way.

Figure 1.3 illustrates how every large .7 density swirl would create smaller .6 density swirls at their edges, that would in turn create .5 density swirls at their edges, creating .4 density swirls at their edges, cascading down to very tiny swirls. This is called the *Cascade Effect*. The larger swirls would eventually become supermassive blackholes. The intermediate and smaller sized swirls, say at .3 density and below, being too weak to sustain themselves would eventually disappear. Their disappearance would leave very large voids that would eventually make up ninety percent of space.

The intersection of the larger swirls would settle into super clusters and clusters of SMBHs. The voids would be bordered by those clusters, groups, and filaments. Within the clusters, the gravitational influence of the blackholes would slow the expansion between them, while the voids would expand more rapidly, creating even larger voids.

Figure 1.4 Depiction of the cosmic web. Credit: Virgo – Millennium Simulation Project/Springel et al. (2005)/Max Planck Institute for Astrophysics.

As Physicist Charles Seife tells us, 'We live in a Swiss cheese universe.'[34] The Cascade Effect is very consistent with the creation of the Cosmic Web of galaxies we see today, as illustrated in the artist's depiction in Figure 1.4.

The cosmic web matches exactly the kind of galactic distribution we would anticipate if the SMBHs started out as large swirls cascading to smaller swirls, with the intermediate swirls eventually disappearing. Based on all the observational evidence just cited, the argument for supermassive blackholes developing prior to star formation is very plausible. If the supermassive blackholes were created out of the dense Penergy of the early universe, it would mean that the particles were *not* created within the first minutes following the big bang but created well after the supermassive blackholes began forming. It also means the two tentative deductions made earlier appear valid.

> Deduction #4: Considering the observational evidence, it is reasonable to conclude that supermassive blackholes were created out of pure energy from the dense Penergy of the early universe, and their development was well under way when the first particles and stars evolved. Additional evidence presented in the following chapters will also support the validity of the first three deductions.

When scientists extrapolate their theories on how the universe will ultimately evolve, they place blackholes as the last things in the universe to exist. This seems appropriate, as it appears they were also the very first things to exist.

> Prediction: Someday a bright young physicist will develop the math to properly describe Penergy density. Once that is accomplished, the ability to create a computer simulation of the early universe will likely confirm that SMBHs can be, and probably were, created by some calculated density of Penergy.

The First Stars

As the SMBH gravitational influence grew, it would have slowed the rate of expansion and rate of Penergy density reduction. This would have allowed the smaller creations that spun into existence to live much longer before disintegrating back into the Penergy medium. It's possible the creations the size of our sun and larger could have existed long enough and had sufficient gravitational influence to draw in some of the smaller creations and early particles surrounding them, giving them enough mass and angular momentum to forestall disintegration, allowing them to eventually turn into our first stars.

THE ORIGIN OF SMBHs & THE COSMIC WEB

* * *

Starting with only two initial conditions: *the universe started from a speck of pure energy,* and *it began to expand,* our initial observations and deductions have allowed us to begin to build a very interesting foundation to the universe. We first concluded that if the initial speck of pure energy (*Penergy*) began immediately expanding and thinning, then it was likely capable of existing in multiple densities. From there we concluded it is unlikely our subatomic particles were created within the first minutes of the big bang as theorized, and that their creation had to await further cosmic expansion and Penergy density reduction.

Since everything in the universe spins, it is a reasonable hypothesis that at some level of density large clumps of Penergy began to swirl on their own accord. This set off a cascade of ever smaller swirls, ultimately ending in only two stable creations: supermassive blackholes (SMBHs) and tightly wound early elementary particles (EEPs).

Since it appears to be a reasonable way for SMBHs to have been created within the time frame calculated for their existence, and since the cascade scenario perfectly reflects the *cosmic web* galactic structure we see in existence today, we concluded that the notions of both Pure Energy SMBHs and the Cascade Effect are very plausible.

Our young universe has gradually evolved through multiple Penergy densities. Its near-infinite density state as a big bang speck has already evolved into medium-high density nascent SMBHs and medium-light density smaller swirls. Next, we will examine two of its light-density states, the evolution of EEPs and its light-density *medium* state, aka, Dark Energy. The gradual thinning density of Penergy may have produced other conditions in our cosmic history that are not yet apparent to us. Perhaps future discoveries and enigmas will be answered by the concept of a constant thinning Penergy density.

Chapter Two
The Origin of Particles & Dark Energy

At this point in our story our expanding universe is comprised of billions of developing supermassive blackholes, billions of disappearing intermediate swirls, and a mass of very tiny, stable swirls, dubbed for now to be early elementary particles (EEPs). The timing of the larger swirls developing into blackholes, and the cascade ending in tightly wound swirls, was apparently just right to spare many of the tiny swirls from being drawn into the developing blackholes. Some of the intermediate sized swirls may have gone that way, however, perhaps feeding larger swirls capable of becoming the first proto stars.

Over the next three chapters we will discuss the four phases of cosmic evolution in the order of their creation: EEPs, interspatial Penergy medium (*aka, dark energy*), Fields, and Forces. Each of these phases naturally evolved from the conditions created by the phase that preceded it. We will begin with a brief description of the first particles to spin into existence but hold off on a discussion of their continued evolution until Chapter Four.

The Penergy Density has thinned substantially since the universe began. The following chart is meant to show how that thinning related to evolutionary events, but it is only a rough idea of how that may have happened and is not meant to be in any way precise.

Cosmic Event	Penergy Density
Big Bang	1.0
Penergy begins to swirl into large independent entities that will become supermassive blackholes.	.7
The swirls cascade into ever smaller and smaller swirls.	.7 - .2
The cascade continues, producing tiny swirls we call early elementary particles (*EEPs*).	.2 - .15
EEPs combine to create the First Generation of composite particles.	.15
The First-Generation phases into the Second Generation.	.10
The Second-Generation phases into the Third Generation.	.05

Figure 2.1 List of evolutionary phases with guesstimated Penergy densities.

Early Elementary Particles

The consequence of emergent blackhole swirls is that the density at the outer edges of those cascading swirls would get thinner and thinner until it was thin enough for things we call Early Elementary Particles (EEPs) to spin into existence. It may seem a stretch to relate the formation of blackholes and particles from the same creation dynamics, but the idea isn't new. String theory (*an alternative theory for the physical make-up of particles*) says blackholes and particles may be manifestations of the same basic entity, namely *strings*. Here I am saying that the same basic entity is Penergy.

The particles that spun directly out of the Penergy cannot be today's Sub-A's without some explanation as to how those particles could have formed so quickly with such complexity. Complexity arises in nature not from a sudden and inexplicable appearance, but from *evolution*. As discussed earlier, Penergy spinning itself directly into those complex particles without an evolutionary foundation just doesn't seem reasonable. We are therefore compelled to conclude that these freshly spun, simple particles are new and different, requiring new names and descriptions.

Inventing new particles is risky for any theory, so proceeding is accomplished with a good deal of trepidation. What follows is not meant to be an exact picture of how the earliest particles evolved but only how they *may* have evolved. Our further construction of the early universe remains as much as possible based on keen observations and good deductions, but we are diverging significantly from current theory. We will be introducing two levels of particle matter below the subatomic level we are familiar with today. Don't be thrown off by these new names and descriptions. This picture of the early universe is unique but remains very reasonable and relatively straight forward. It is simply a story about little particles that combine and evolve into more complex particles. As you will see, it is a story far more plausible than one about particles that simply pop into existence.

The first stable and dominant particles that spun into existence were truly elementary. It means that our electrons and quarks must be composite particles built from these new particles. Although there is no solid evidence for it, the argument for composite particles is reasonable and compelling. Consequently, we will proceed with the notion that the first, early elementary particles are new to us and were comprised of nothing more than tightly wound Penergy.

By spinning into existence, these particles would presumably look like tiny tornadoes, so let's call them Torons, or just Tors for short. Everything we say about them will be new and not about any existing Model, so let's call all the related observations and deductions part of a *Tor Model*.

The creation of these new particles ushered in new dimensions to our universe. The creation of distinct bits of particle matter with a finite distance between them implies *space*. Particles flying away from each other at a finite speed implies *time*. These particles were born into an ever-expanding cloud of Penergy; a thinning texture of the interspatial medium we often refer to as *spacetime*. It is called spacetime because as it turns out the *space* and the *time* are relative to each other, but spacetime is still nothing more than our thin-density Penergy medium. Space and time are relative because time changes whenever space (Penergy) changes. We will examine in depth why space and time are relative in Chapter Seven.

We've learned from observations at particle accelerators and colliders that a high energy particle called a photon can decay into particles of matter. This decay process results in a pair of particles, one spinning to the left and one to the right, which fly off in opposite directions. Whenever particles are formed through the decay of a large particle created in the Large Hadron Collider (LHC), the decay always results in the creation of a particle pair.[1] From those observations it is reasonable to conclude, as the Standard Model does, that the original creation of particles out of Penergy resulted in the creation of opposite-spinning *particle pairs*.

Those tornado-like Tors would not be loosely wound like Kansas tornados. Being created out of Penergy they would be very tightly wound, giving them size, location, mass, and a solid-like composition. Being tightly wound they would not be shaped like an elongated Kansas tornado, but more likely in the shape of a sphere, as we find spinning stars, planets, and blackholes. Created in pairs having opposite spins means they have a spin axis and an orientation of top and bottom, one might refer to as a nose and a tail, respectively. Their spin direction is determined by looking at their tail, down their spin axis. The point of these observations will be important as we discuss what capabilities and attributes these particles possess.

Particle Generations

Once stable Tors evolved, they would have begun combining, creating the first generation of composite particles. Today we observe three known generations of particles, two of which are unstable and quickly decay into lighter versions of themselves. Each generation is substantially lighter in mass than the previous generation.

No one knows why there are three generations. The Tor Model speculates that the first generation may have come into existence at say .15 Penergy density. As the Penergy medium thinned, apparently those particles could not exist in the thinning density and disintegrated into lighter versions of themselves representing the second generation of particles.

That second generation having come about at say .10 density apparently could not continue to exist in that state as the surrounding Penergy density thinned further. They went through another phase transition and disintegrated into say the .05 density third generation particles we see today. Apparently those .05 density particles, the core elements of our Sub-A's, are quite stable in the existing Penergy density medium and will likely stay that way. Photons went through the same process, but we don't recognize their phases as generations. Their evolution will be discussed in Chapter Six.

The stated values of the densities at which those phase transitions took place are of course only guesses and the actual densities could be quite different. The first and second generation of particles will now only appear when there is a knot of Penergy dense enough to bring them into existence, which happens in our particle accelerators and colliders. Of course, they are in existence only momentarily, not being able to sustain themselves in the now prevailing very thin Penergy medium.

Again, those first generations of composite particles were not spun directly out of the Penergy but were composites of the Tor particle. It is from those Tors that each generation of Sub-A's evolved. There is only modest evidence for the existence of Tors, but they make sense for several reasons, which we shall examine.

Particle Attributes

We don't know the specific attributes of these early elementary particles, but we might gain a picture of some of their attributes by making observations about the Sub-A's into which they evolved. The natural tendency for Penergy is to swirl and spin, which is how Tors came into existence. Consequently, we can conclude that Tors spin on an axis. Spin is very important to the workings of Penergy. It is how it changes from a formless cloud into a defined entity with mass, momentum, and other important characteristics we shall discuss in due course. Tors are the only particles spinning on their own axis, which will be important later, as well. Spin is at the core of everything in existence, making it one of the most important dynamics in the universe.

We have observed our Sub-A's to be quite resilient. They can be moderately knocked about like billiard balls without losing their composure or identity. Since Sub-A's and Tors are made from the same state of Penergy, we can conclude that Tors are equally resilient. Perhaps because they are more elementary than the Sub-A's, they may be even more resilient.

This is important because resiliency may be the reason we cannot yet breakdown Sub-A's into deeper levels of constituent particles. The deeper we go, the smaller, more compact, and more resilient those particles are likely to be. We may never see an early elementary particle and may forever know them only theoretically.

We have observed that Sub-A's are known to have *fields*. A field is an area of influence immediately surrounding the particle. Since we have concluded that Tors are tightly wound, fast-spinning particles like Sub A's, it is reasonable to conclude they too have an area of influence surrounding them. Fields are very important in the reconstruction of our universe, so they will be discussed in depth in the next chapter.

As mentioned earlier, when matter particles are created out of Penergy they are created in pairs with opposite spins. Scientists label the pair of particles *matter* and *anti-matter*. The two particles have the same mass and are otherwise alike in every way except for their direction of spin and an attribute scientists call *charge*. The particles are said to be a *mirror image* of each other. Only Tors have their charge type associated with a particular direction of spin. Composite particles can spin in either direction irrespective of their charge.

Scientists have observed that when matter and anti-matter particles meet, they annihilate each other, turning back into a state of pure energy. In the early particle creation era, there was a great deal of particle annihilation and photon production. This would have heated the environment significantly, creating a very hot frenzy of activity. If matter and anti-matter were created in pairs simultaneously and therefore in equal numbers in the early universe, why is there predominately only matter in the universe today? What happened to the anti-matter?

The Mystery of the Missing Anti-Matter

No one knows for sure where the anti-matter went. It is one of the universe's biggest mysteries. The current theory speculates that due to an imbalance in the creation/annihilation process in the early universe, out of every billion annihilations there was an imbalance leaving a single matter particle.[2] If this phenomenon happened over and over, it would leave the universe with an abundance of matter particles with few anti-matter particles, which is what we observe today.

That scenario is difficult to believe given the enormous number of matter particles in the universe and the low probability of the same imbalance happening over and over for all known particles. It's also a stretch to imagine the timing of it, since all the matter particle imbalances would have to have occurred within the first seconds after the big bang. That is a lot of annihilations, creations, and imbalances to have occurred within a few seconds. The annihilation imbalance theory stretches plausibility too far. There must be a better explanation.

There are likely other theories for the missing anti-matter, but the least complicated and most straight forward theory is that the anti-matter isn't actually missing. Matter and anti-matter are said to have opposite electric charge values and the universe is said to be charge neutral, meaning equal in charge value.[3] If that is so, for the universe to be charge neutral, the anti-matter must still exist. But if the anti-matter still exists, where is it?

The Case of the Anti-Missing Anti-Matter?

The most straight forward answer may not seem feasible at first glance, but it develops into a reasonable hypothesis. If the Sub-A's we know to exist – the quarks, electrons, and neutrinos – are composites of smaller elementary particles, then the missing anti-matter could be a part of the constituent particles that make up the Sub-A's. Anti-matter could be an integral part of the matter we see all around us. Assuming for the moment that is even possible, is there any observable evidence for the hidden presence of anti-matter?

As mentioned earlier, particles of matter and anti-matter are mirror images of each other, differing only in the direction of their spin and charge. Charge can be either negative or positive; attributes ascribed to matter and anti-matter respectively. To find anti-matter therefore we only need to hunt down particles carrying a charge opposite of that of matter. But we know of those particles already. Electrons have negative charge and protons have positive charge. Up quarks have positive charge and down quarks have negative charge.

If matter and anti-matter are known to carry opposite charges, and we find both types of charge existing in the particles around us, it suggests each type of matter exists respectively within those particles. If down quarks are negatively charged and referred to as matter, it makes sense that up quarks that are positively charged contain at least some anti-matter. These are bold ideas with some obstacles to overcome but given no better explanation for solving the anti-matter mystery, they are worth examining.

If our Sub-A's possess both types of matter it would mean our Sub-A's are definitely composite particles and that composite Sub-A's could solve the missing anti-matter mystery. There are good arguments for the existence of composite particles, and they have many other benefits as we shall see. We shall proceed with both notions: that the first particles to spin into existence were true elementary particles from which Sub-A's evolved, and their constituent particles might include anti-matter.

With these notions in mind, let's continue with our reconstruction project. First, we must consider our biggest obstacle: how does one construct composite particles by mixing matter and anti-matter together without them annihilating? As my grandmother would say, 'That problem's a doozy.'

Spin Signature

Annihilation occurs between colliding particles when the two are mirror images of each other as are all matter/anti-matter pairs. An electron and its mirror image, the anti-electron, will annihilate, but an electron and an anti-muon (*a second-generation version of an anti-electron*) will not annihilate.[4] Neither will an electron and anti-proton.[5] In each case the two particles have opposite spin and charge, but apparently the mass difference between the respective particles precludes annihilation.

In another example, a subatomic particle called a meson is made up of a quark and an anti-quark. The two quarks can co-exist together without annihilation because of a characteristic called *color* that differs between the two. Apparently, the slight difference between all those examples is enough to preclude the usual matter/anti-matter annihilation.

It seems that to annihilate, the two particles must be *perfect* mirror-images of each other, and a deviation in that perfection will preclude annihilation. As Physicist Bruce Schumm tells us, "…matter/antimatter annihilation takes place if and only if the matter particle comes into contact with its exact anti-matter counterpart…"[6] But how does an electron know whether it is in the presence of a positron? There must be something inherent in each particle that identifies its mirror-image.

Particles have few primary traits – mass, spin, and charge. For mirror-image particles to annihilate, we know based on the examples above that the mass must be the same between the particles, and we know the charge must be opposite. That leaves only spin as the inherent, identifying trait. It would mean that each particle type has its own distinct spin, or *spin signature*, and only particles with the exact mirror-image spin signature can annihilate.

The difference in spin signature preventing annihilation might be something subtle such as a difference in the wavelength (*think energy level*) between the particle and anti-particle. Physicist Michio Kaku reports that in a very cold trap to store anti-protons in a magnetic field, the wavelength of the cold anti-protons would be much longer than the wavelength of the protons of the atoms in the container walls, so the anti-protons would reflect off the walls without annihilation.[7] Here are touching, mirror-image particles, but their respective energy wavelengths apparently constitute a significant deviation creating slightly different spin signatures, thus no annihilation.

> Deduction #5: Given that particles with opposite spin and charge don't annihilate when their masses or other traits are different, we can conclude there is an inherent trait that allows particles to identify their perfect mirror-image counterparts. We have dubbed that trait *spin signature* and concluded that only particles with a mirror-image spin signature annihilate.

> Prediction: Experiments involving different configurations or states of anti-matter will confirm the validity of spin-signature, or a similar theory, paving the way for a complete, composite particle theory.

We have identified the first surviving particles as particle pairs – left and right spinning, mirror imaged, tiny tornado-like, tightly wound spheres of Penergy we are calling Tors. Those Tors would eventually combine to create matter, which will be taken up starting in Chapter Five. Before that phenomenon took place, however, there were other phases of evolution the universe underwent. As the universe expanded, the Penergy medium continued to thin past the particle creation phase, and the universe entered the next phase of evolutionary development – the creation of the thin Penergy Medium, aka Dark Energy.

The Origin of Dark Energy

Since Astronomer Edwin Hubble in 1929 observed that galaxies were moving away from each other concluding the universe was expanding, scientists have speculated on what is causing that expansion. The consensus is that it must be in a form of energy, but they know of no existing energy that could do that or how it can even exist.[8] The mysterious energy has been dubbed Dark Energy.

The Standard Model says everything was created from the speck of Penergy within the first second of the Big Bang, with most of the energy in the form of *light* and the rest in the form of fundamental particles, mainly protons, neutrons, electrons, and positrons.[9] The Standard Model seems to assume that in the creation of those particles of light and matter, all the Penergy in the original speck was consumed. We have already concluded that the notion of complex particles popping into existence within the first second is unreasonable, so the later assumption that all the Penergy was consumed within that second is also unreasonable.

Let's see what more we can learn from the initial conditions we started with: the universe began as a speck of very-dense Penergy and for reasons unknown began expanding. We can infer that Penergy when at some density less than its very-high-density NIB-state, naturally expands. As the universe expanded and the Penergy thinned, we have concluded it eventually reached a density that allowed particles to come into existence. There is no reason to believe that all the Penergy was consumed in that particle creation epoch. That would mean that an even thinner density volume of Penergy could still exist, pervading all of space.

That prospect is not only plausible but seems obvious since according to quantum theory virtual particles (*think particles that quickly pop in and out of existence*) can be momentarily created by borrowing energy from the vacuum of space. What other energy could those virtual particles borrow from if not the same energy from which all real particles and everything else in the universe is made? For virtual particles made of Penergy to pop into existence, that Penergy must be present in the vacuum.

Mario Livio points out in his book, *The Accelerating Universe*, that particle-antiparticle pairs can also borrow energy from the vacuum, and those particles immediately annihilate themselves back into the vacuum. He characterizes the vacuum as bubbling with such virtual pairs.[10] According to Physicist Lisa Randall, when a proton and anti-proton annihilate, they turn back into pure energy.[11] Again, if known particle pairs are being momentarily produced out of and disappear into the vacuum of space, that vacuum must contain the requisite energy to produce those pairs. The creation of particle pairs implies that the energy of the so-called vacuum of space is the same energy from which real particles are created.

We can conclude from all of this that the vacuum of space is not actually a vacuum but is filled with an energy capable of producing virtual particles and particle pairs. The only energy capable of doing that as far as we know is the same energy from which all other particles are made - Penergy.

If all the Penergy was not consumed creating particles in the early universe, it would leave the residual Penergy growing even thinner as the universe expanded. Though thin, apparently the fluctuations in the existing Penergy density can still allow particles to momentarily come into existence. This suggests the current Penergy density is just below the particle-creation density threshold but is dense enough for fluctuations in that density to momentarily create particles.

This remaining very-thin Penergy could now simply serve as our interspatial medium and be the source of energy that is continuing to expand the universe. Dark Energy could be nothing more than a very-thin-density version of the Penergy that started the universe. The same Penergy from which everything is made and from which virtual particles and pairs borrow, and into which they disappear.

The Interspatial Medium

We know that the interspatial medium consists of something, because scientists report evidence that the spin of the earth [*and all other large bodies*] is causing frame dragging, meaning the earth's spin is dragging the adjacent spacetime with it.[12] Possibly the actual substance being dragged is the thin Penergy medium, aka dark energy.

Further evidence our interspatial medium is a fluid-like substance and not a vacuum comes from evidence of gravitational waves, which were predicted from Einstein's theory of General Relativity. Gravitational waves are disturbances in the interspatial medium caused by cataclysmic events such as the crashing together of neutron stars or black holes. Those disturbances create waves that propagate across space. Scientists have built the very sensitive equipment necessary for their detection and in 2016 reported that the first gravitational waves were detected.

Gravitational waves stretch and distort along their path; a circle becomes an ellipse, and a square becomes a rectangle.[13] According to Professors Jorge Cham and Daniel Whiteson, this kind of behavior can only happen if space has a certain physical nature to it.[14] There must be something doing the waving and stretching.

Scientists once thought space contained a substance they called *ether*, but the presence of such an ether was disproven in the famous 1887 Michelson-Morley experiment. The presence of the Penergy may be difficult to prove, as well. It is not a physical substance as we know it. It is not made from particles that would have mass or possess other particle-like characteristics that we can detect. Physicist Laura Mersini-Houghton tells us, "Dark energy behaves like an ether: its energy mysteriously diffuses out of a vacuum and yet it permeates every speck of the universe and pervades the very fabric of space-time."[15] Penergy may not be directly detectable, but we might be able to infer its presence from its relationship to particles and virtual particles, as discussed above.

Einstein's theory of General Relativity taught us that gravity is the curvature of spacetime, which is often referred to as a fabric. That notion implies that the fabric of spacetime is something that can be curved and distorted since that is the only reasonable way it could influence the path of passing matter. Astrophysicist Paul Sutter tells us that General Relativity taught us that spacetime itself is a dynamic, living, breathing, physical object.[16]

We also know that space can be made into waves and can be expanded. All these characteristics make more sense if they apply to the existence of *something*. That something may be difficult to detect because it is not like anything else we encounter, but there is little doubt the vacuum of space is no vacuum but is filled with some form of energy. The energy that makes most sense is a thin-density version of the original Penergy from which the universe began its expansion. Based on the above observations we can make our next deduction.

> Deduction #6: If the original Penergy speck was capable of multiple densities and was not immediately consumed in particle creation, some thin-density Penergy may still exist today comprising our interspatial medium. If everything in the universe is made of Penergy, then it stands to reason that that very thin-density Penergy is the "Dark Energy" that has been expanding the universe since its inception.

* * *

In this chapter we traced the cascade of Penergy swirls down to the particle level, and then beyond that point to the layer of thin density Penergy that currently permeates the universe. Since those early particles that spun out of the Penergy were not our Sub-A's but entirely different particles, we named them Tors and created what will become the Tor Model.

The energy permeating the universe that scientists call Dark Energy sounds very much like the Penergy that started the universe. It is the same energy from which virtual particles and particle pairs are created and into which they disappear. We concluded that Dark Energy and Penergy are one and the same.

We now have Penergy existing initially in a very-high *NIB-density* state, in a high *SMBH-density* state, in a light *particle-density* state, and in a very-light *dark-energy-density* state. We have built the first layer of our universe upon the foundation of supermassive blackholes, early elementary particles, and dark energy, which we shall refer to henceforth as the interspatial Penergy medium. They represent the first true building blocks to our physical universe. Importantly, they naturally evolved from simple initial conditions: a speck of pure energy that naturally began expanding.

Bringing new particles into theoretical existence is risky for any model because we have no solid evidence for their existence and must rely on circumstantial evidence and reason. But their origin by spinning into existence and evolving into today's subatomic particles makes more sense than having them simply pop into existence without foundation or explanation. Composite particles answer the mystery of the missing anti-matter and provide a better origin for fields and forces, as we shall see.

We next turn to the first *relationship* to be established in the universe, opening a whole new era of creative evolutionary advancement – the origin of fields.

Chapter Three
The Origin of Fields

The initial speck of Penergy has expanded substantially and has lost most of its density, thinning past the point it could naturally spin more particles. The gradual thinning density left in its wake SMBHs, tiny particles, and a thin, pervasive medium. The Cascade Effect created weak intermediate sized swirls, but they are steadily disappearing. It is unknown how long it took for SMBHs and particles to come into existence, but it was likely a very long time, which allowed the universe to grow substantially.

With the creation of SMBHs, particles, and the interspatial Penergy Medium, we have laid the foundation to our universe. These three stable densities of Penergy represent the first phase of cosmic evolution. Everything we know to exist in the universe, including time, was built from *relationships* between these three densities. That statement is so monumental it is worth repeating. *Everything we know to exist in the universe, including time, was built from relationships between these three densities.*

Mathematics is very important to our understanding of the universe. One of the reasons it describes our universe so well is that mathematical equations describe relationships, and relationships and symmetries are at the core of our evolutionary structure. We next examine the very first relationship to come into existence: the relationship between particles and the Penergy Medium. That relationship produced *fields*.

Particle Fields

A field for our purposes is an area of space containing a common, measurable characteristic. The field can be large, such as an entire weather front, wherein each point in space has a measurable temperature. The overall area so measured would be considered a *temperature field*. A field can be small, such as the field surrounding a particle possessing an electric charge. At each point in that space there is a measurable charge value and the overall space so measured would be a *charge field,* a.k.a., *electric field*.

In the Tor Model, a particle's field emanates from the spinning gyrations of the particle that affects the surrounding Penergy medium creating a bubble-like area immediately surrounding the particle. That whirling field portrays the particle's characteristics and provides the means for particles to communicate and interact. Particle fields do not have precise boundaries since the measurable values of influence gradually dissipate with distance from the particle. They do have a practical boundary, which is the margin at which the field has an insignificant influence on other particles.

Opinions differ as to how fields are created. Do fields create particles, or do particles create fields? In Quantum Field Theory (QFT) fields create particles. Those fields are ubiquitous throughout the universe and for every particle there is a specific field. QFT says particles have no size, take up no space, and are only manifestations of a field.[1] A quark is simply an excited state of the ubiquitous quark field.

QFT and the Tor Model do have similarities in their view of the universe. QFT recognizes that particles and fields are associated, but it is the fields that are ubiquitous throughout the universe and from which everything else is made.[2] In the

Tor Model, Penergy is ubiquitous throughout the universe and from which everything else is made.

In QFT, the ubiquitous fields are the elements that play the role of a communications network between particles.[3] In the Tor Model, Penergy is the element that plays the role of a communications network between particles. Everything discussed in the prior sections suggests Penergy is real. There is no evidence ubiquitous fields are real, outside of QFT equations.

QFT is another mathematical edifice that describes a reality difficult to believe. Physicist George Musser tells us physicists struggle to understand what quantum field theory is telling them about the world.[4] Chemical Physicist Michael Munowitz says that, "a quantized field at first blush is a monstrosity, a mathematical absurdity that makes no sense at all."[5]

Physicist Sean Carroll tells us, 'The Core Theory [*a.k.a. Quantum Field Theory*] is not the most elegant concoction that has ever been dreamed up in the mind of a physicist, but it's been spectacularly successful at accounting for every experiment ever performed in a laboratory here on earth.'[6] Quantum math is torturous, but it works. But that does not mean QFT's *interpretation* of how that math reflects on reality is correct.

QFT might be mathematically accurate, but it is more feasible to have fields arise naturally due to the relationship between particles and the Penergy medium, rather than due to a scheme created to fill a need but without a good foundation. It is more compelling to consider the field as simply a local disturbance in the Penergy medium due to the dynamics of a particle's vibrating presence. Let's examine what that might look like.

The Attributes of Local Fields

Scientists have observed that particles such as protons and neutrons vibrate with their own specific frequency. In the Tor Model that vibration is the spinning gyrations of the particle affecting the surrounding Penergy medium. As the spinning particle moves forward, it corkscrews through the Penergy medium creating waves. Doing so makes a disturbance that pulsates with typical wave-like features, such as wavelength, intensity, and frequency. This point may be at the heart of the issue as to how a particle can exhibit both point-like and wave-like characteristics. The matter is commonly referred to as the wave-particle duality issue and is a core feature at the foundation of Quantum Theory. We will examine the duality issue in detail shortly.

Particles affect the Penergy medium in proportion to the disturbance they make, making the field size generally in proportion to their makeup and spin characteristics. These observations imply that the field of even a single particle is much bigger than the particle itself, surrounding the particle like a cocoon or bubble. As previously observed, the bubble does not have a specific boundary; the field gradually fades into a practical boundary.

Physicists have espoused the idea of particles creating wave-like fields. Physicist Michio Kaku tells us that the photon particle creates both an electric and magnetic field surrounding it, and the fields are shaped like waves and obey Maxwell's equations (*summarizing the electric and magnetic relationship*).[7] That observation suggests that it is the field that is doing the waving. Dr. Kaku may not have intended to make that specific point, but his word choice certainly suggests it.

We have observed from electric fields that the presence of many of the same particles have the cumulative field strength of all the particles involved. As different particles combine, their fields combine forming a new field holding the characteristics of the new combination of particles, e.g., a proton's field is much different than the fields of the individual quarks making up the proton.

As particle/bodies combine and gain more mass, their fields, though possibly more complex, are not necessarily proportionally larger. As the particle/body scales up in mass, the movement of the particle/body within its field is gradually less determined by the subtleties within its environment. Once it reaches a scale where the particle is seldom making responsive moves within its field, it leaves the quantum world and joins the macro-world. In the macro-world the attributes of the particle/body can be measured with certainty, no longer needing Quantum Mechanics, though it remains quite accurate if used. Fields remain an attribute of larger bodies, including humans who refer to their personal fields as auras.

Quantum Theory recognizes particles having attributes, but there is a controversy as to where some of those attributes reside and when they come into existence. One would think that a particle has a certain attribute, or it doesn't. And it should not matter whether that attribute has a fixed value such as charge, or a variable value such as its direction of spin. Apparently in quantum mechanics that is not the case. Philip Ball in his book *Beyond Weird,* tells us that quantum mechanics does not recognize particle variables that can be assigned in advance even though they appear to acquire those values randomly through the act of measurement.[8]

Consequently, those variable characteristics do not appear in quantum theory's mathematical descriptions of the particles. In other words, despite knowing that electrons are either spin-up or spin-down, its spin direction *does not exist* until it is measured and determined. Those variables that particles can possess were deemed hidden from view and became known as hidden variables. Whether variables exist that are capable of being pre-set prior to measurement is an open issue and is at the heart of a phenomenon called *entanglement*, which is another core element supporting Quantum Theory and will be discussed at length in Chapter Eight.

In the Tor Model, particles have many attributes or variables, all of which are real, whether hidden or not. Some of these attributes are represented in their fields, as listed below. There may be more.

- Since all particles vibrate at their own peculiar intensity and frequency that is reflected in the particle's field, if one could *read* the field, one could identify the particle by those attributes.
- A photon's state includes some value for its polarization. Polarization is the direction in which the photon is determined to be doing its waving. Since that which is waving is the photon's field, then its polarization must be said to be an attribute of its field.
- Since a field has a wave-like attribute, and the frequency of the wave is a function of the particle's energy, one can say that the field is also a reflection of the particle's energy.
- Some believe the particle's field even allows for communication. Lynne McTaggart tells us in her book *The Field,* 'These subatomic waves or particles not only know about each other, but also are highly interlinked by bands of common electromagnetic fields, so that they can communicate together.'[9]

Notwithstanding the approach taken by Quantum Theory, it would seem those variable characteristics are real and a part of the particle, whether measured or not. Further evidence for local (*personal*) particle fields will be discussed next.

Deduction#7: Given the cumulative evidence for the existence of the Penergy medium; and given that the Penergy medium can serve in the same capacity supporting particle fields as QFT's universal fields; and given the absence of evidence for the existence of universal fields except mathematically; we can conclude that the source of particle fields is more likely a disturbance in the Penergy medium caused by the spin characteristics of individual particles. Additional support for this deduction will be revealed subsequently.

A Particle's Personal Field

In the Tor Model, fields are not ubiquitous throughout the universe, but are personal to individual particles and groups of particles making up any size body. The field was referred to earlier as a bubble around the particle but there is no physical bubble. The bubble-like cocoon is used only as a visual image for the presence of a field. We can't see individual particle fields, but we know enough about how they affect other particles to have a vision of their presence. There is no dispute as to particles having fields, only the source of those fields.

The argument for a particle having a local, bubble-like field is found in our understanding of *particle location*. According to Quantum Theory as told by Tim James in his book *Fundamental*, we cannot know a particle's precise location until we measure it, and that particle locations are determined by waves of probability.[10] He points out that particles can be pushed through the same experiment over and over and come out in a different place each time. Quantum Theory ascribes this phenomenon to the particle also being a wave, which gives it an indeterminate location. Given the current belief in particle duality, this notion is justified, but not necessarily accurate.

Scientists never see particles as waves, only as point-like particles.[11] This is important so allow me to repeat it. *Scientists never see particles as waves, only as point-like particles*. In other words, there is no physical evidence that a particle is also a wave. Particles are believed to be waves based on the interpretation of experimental outcomes, such as in the double-slit experiment (*discussed in the next section*) and from the experience of a repeated experiment resulting in different location outcomes. These experimental outcomes seem counterintuitive and weird, fostering a wave-like interpretation, but they can be explained by visualizing a particle that is producing its own individual, local field within the Penergy medium.

In the Tor Model, a particle is not a wave. It can seem to behave like a wave because its field, created by the particle's spinning gyrations disturbing the Penergy medium, has the characteristics of a wave. This idea is confirmed by James Geach who tells us in his recent book, *Five Photons*, "... for a charged particle surrounded by an electrostatic field, any oscillation of the particle will also oscillate the field."[12] In terms of a personal field, as the particle moves, it constantly affects the Penergy medium creating a bubble of disturbed Penergy around itself, thus the oscillations of the particle will also oscillate the field.

The bubble of disturbed Penergy is not fixed. It is very fluid, and its dimensions, shape, and behavior are all subject to minute changes depending on what the particle is responding to at any given moment. The Penergy medium is a frenzy of activity due to the volume of particles, forces, gravitational waves, energy fluctuations, etc. Consequently, the bubble can be distorted depending on what condition the particle is responding to.

Because a particle is constantly affecting the Penergy medium, it is in essence constantly creating a new bubble of disturbed Penergy as it moves. The on-going creation of the bubble causes a lag time between the location of the particle and the location of the practical border of its field. The bubble is constantly shifting around the particle as the particle is moving about within its minute environment.

Again, the particle's gyrations create vibrations that change the state of the Penergy, creating a distinct environment that defines the bubble-like field. Within that bubble environment the particle is free to roam while at the same time being seemingly confined by the bubble. While the particle is creating the bubble, the bubble is giving the particle a safe and defined place in which to reside. It is an extension of the particle, like a barrier protecting the particle - a first line of defense and communication when encountering another particle. The practical result is that the particle can be found anywhere within its bubble-field, with its precise location determined by mathematical probabilities. Those probabilities are accurate using a wave-like theory because the particle's movements, its field, and the Penergy medium are all behaving in a wave-like manner.

Given the above description of a particle's personal field and the idea that fields are without boundaries, it is understandable why quantum theory posits particles could be anywhere until they are measured or detected. Some fields such as the electric field may be without a boundary technically, but the density of the field falls off so quickly that its practical boundary, where it has an insignificant influence on other particles, is not terribly far from the particle itself. Therefore, the probability of finding the particle far away from the center of its field becomes remote. Again, at any moment the particle might be found anywhere within its practically bounded field, which explains why one would obtain different location-results with an identically repeated experiment.

Another instance of particle behavior suggesting it is a wave has to do with its apparent capacity to tunnel into places to which it theoretically should be incapable of moving. A particle shot at the edge of a wall will usually bounce back, as we would expect a tennis ball to do. But occasionally the particle can be experimentally found on the other side of the wall. This is known as *particle tunneling*.

In quantum theory the interpretation of the tunneling phenomenon is that the particle is a wave that occasionally diffracts (*bends*) around the edge of the wall, as waves are known to do. Its bending around the edge of the wall allows the particle to show up on the other side. This is of course only possible if the particle is a wave.

In the Tor Model, it is the particle's field that is diffracting around the edge of the wall. Since the particle is seldom located at the edge of its field, it is seldom that it is within the portion of the field that is being diffracted, so it seldom appears on the other side. Occasionally, however, the particle is at the edge of its field that is being diffracted and it shows up on the other side. We can learn more about wave-particle duality and diffraction from the famous double-slit experiment discussed next.

The Double-Slit Experiment

One of the cornerstones of Quantum Theory is the idea that particles are both a point-like particle and a wave, giving their behavior a great deal of uncertainty and possibly meaning particles are not real at all. True enough, particles exhibit point-like particle attributes in scattering experiments where they ricochet off each other. And they exhibit wave-like attributes in experiments like those involving electron beams striking crystalline solids in which the electrons exhibit diffraction and interference, both attributes found only in waves.[13] But perhaps there is a plausible explanation for this apparent wave-particle duality.

Quantum particles can appear to be both point-like or wave-like depending on how one has chosen to detect or measure them. For physicists the proof of the wave-particle duality comes in the way of the double-slit experiment. When a wave is washed against a thin wall having two closely placed vertical slits, the slits will cause the passing wave to diffract (*bend around the edges*) as it passes through. The semicircular waves emerging from the two slits will interact with each other because the wavefronts will start to cross each other's path. The diffracted waves from the two slits interfere with each other both constructively and destructively, creating crests and troughs that show up vividly on a detection screen set just beyond the wall. This setup is known as the double-slit experiment. For a complete description of the experiment, see Philip Ball's *Beyond Weird*.[14]

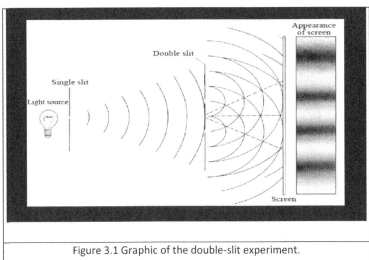

Figure 3.1 Graphic of the double-slit experiment.
(Source: Penn State University Physics)

When sending either photons or electrons through a two-slit setup the detection screen will show the wave-like pattern of crests and troughs, as shown on the right side of Figure 3.1. The conclusion drawn from these experiments is that the individual particles are passing through *both slits at once*, interfering with themselves, and therefore are not only acting like waves, but *are* waves. They originate as particles, and end as particles at the detection screen, but in between they act like waves.[15]

The conclusion being drawn is understandable, but it may not be correct. One can also imagine the same experimental outcome with each particle passing through only one slit but its *field* passing through both slits. It might be the field that is being divided and diffracted. The wave-like appearance of particle locations on the detection screen would be due to the particle itself being somewhere inside the divided and diffracted field. The particle would strike the screen before the two fields could recombine. Its exact location would be consistent with the probability of finding it greater in the constructive waves and lesser in the diminutive destructive waves, hence the detected interference pattern.

It is also noted that in a separate experiment if you place a detection device at each slit to determine which slit each electron actually went through, the interference pattern does not show up on the detection screen. The reason for this may be that the detection device caused the particle's field to be significantly disturbed and effectively collapsed at the entry to the slit, thus the field did not go through both slits and therefore could not produce the interference pattern. This too suggests that in the first experiment it is the particle's field, and not the particle, that is passing through both slits.

Particle Field Collapse

Let's examine what is meant by particle field collapse. Recall that in the Tor Model, the spinning undulation of the particle disturbs the surrounding Penergy medium creating a tension in the medium that defines the particle's field. The particle will always produce a field, but that field does not have a precise boundary. In circumstances where two or more particle fields are forced together, the fields will overlap, which is not a problem until the particles are forced very close to each other. Because two different particles have different tension values, spin rates, and other spin characteristics, closely overlapping fields will create an area of turbulence that effectively eliminates or collapses the usability of that portion of the field.

A detection device at each slit could cause the fields of the respective particles to be disturbed to the point of creating a turbulence that effectively collapses much of the field. Consequently, the two halves of the field cannot effectively interfere with each other, and no interference pattern is displayed on the detection screen.

The Pilot-Wave Theory

Again, particles are always observed as point-like particles, never as waves. In the Tor Model this is because they *are* only point-like particles, but their existence inside the bubble-like cocoon of their pulsating field makes them behave like waves. Physicist Louis de Broglie in the 1920's, and David Bohm in the 1950's, appear to have nearly figured that out. They advanced a theory that point-like particles are guided by, or riding on, *pilot-waves.* The pilot-wave theory describes everything quantum mechanics does and explains a good deal more.[16] The theory assures that particles are both real and present whether observed or not, in opposition to then popular arguments to the contrary.[17]

Bohm believed particles are being pushed around by these invisible guide waves, thus would appear to move in wave-like trajectories.[18] The photons in the double-slit experiment were likewise surfing on pilot waves and it was the pilot wave passing through both slits that caused the interference pattern on the detection screen.[19] Evidently Bohm's mathematics was good, deviating little from current quantum physics, but he could never prove the existence of pilot waves. Initial experiments looked promising but could not be duplicated. The de Broglie-Bohm theory retains a small following, but the theory remains outside the mainstream of quantum physics.[20]

Bohm thought the pilot wave was a vibration in some pervasive and sensitive field he called the *quantum potential*. That description sounds very much like our Penergy medium. According to Physicist Philip Ball there is nothing obviously impossible about the idea of a quantum potential, it's just that there is no evidence for it.[21] It is quite possible the quantum potential Bohm had in mind was in fact our Penergy medium.

> Deduction#8: Given the presence of the fluid-like Penergy medium and the spinning particles within it producing personal fields; and given that the Double-Slit experiment can be reasonably interpreted as the particle's field passing through both slits being diffracted and interfering with itself; we can conclude that particles are not waves but simply behave with wave-like features.

▫ Prediction: A special prize may await the young physicist who extrapolates Bohm's mathematics for pilot waves into the physics of personal particle fields, which may explain the double-slit interference pattern accordingly and resolve the issue of wave-particle duality once and for all. That prize might be shared with the equally enterprising young physicist that devises and carries out the experiment that proves it is the particle's field that is passing through both slits and being diffracted.

* * *

In this chapter we developed the idea of a particle's field being created by its spinning gyrations affecting the surrounding Penergy medium. We advanced the idea that because the particle can be found anywhere within that bubble-like field depending on what the particle is responding to in its minute environment, finding the particle in any specific area of the field can only be determined by mathematical probability, which is why quantum mechanics works so well.

We looked at the double-slit experiment and determined that it was the particle's personal field going through both slits and diffracting rather than simply the particle itself. We then looked at the Pilot-wave Theory that describes quite well both physically and mathematically our theory of personal particle fields existing in a Penergy medium.

The field, a relationship between particles and the pervading Penergy medium, represents the first significant relationship to have evolved in the universe. The next significant relationship to come into existence was between the particle fields themselves. That relationship governs how particles interact and is called a *force,* which we examine next.

THE ORIGIN OF FIELDS

Chapter Four
The Origin of Forces

According to the Standard Model, particles interact through one of four forces: electromagnetic, strong, weak, or gravitational. The Standard Model posits those forces popping into existence within the first second of the big bang and being effectuated by the exchange of virtual bosons (*undetected theoretical particles that pop in and out of existence*). Particle interaction is governed by certain rules that describe the four forces.

According to Mathematician Milo Beckman, "The exact rules for particle interaction in the Standard Model are absurd.... The calculations involve continuum-sums and imaginary numbers and coupling constants and all sorts of ridiculous math..."[1] Physicist Michio Kaku tells us the Standard Model was created by splicing together by hand the theories that described the various forces, so the resulting theory is a patchwork. He reports that one physicist compared it to taping a platypus, an aardvark, and a whale together and declaring it to be nature's most elegant creature.[2]

Make no mistake, the Standard Model's equations describing particle interactions produce accurate predictions. But Dr. Beckman's comments and Dr. Kaku's imagery are good reminders the Standard Model often makes the math work, while creating a questionable *interpretation* of the underlying phenomena.

The Tor Model does not allow anything to inexplicably pop into existence (*rule #2*) and therefore sees force as something that *evolved* out of the conditions of the early universe. Let's examine the idea of force in each model, using the electric force as an example.

The Theoretical Basis for Force

Force According to the Standard Model

In the Standard Model the electric force is conveyed by exchanging virtual *photons* between charged particles. A photon exchange between two like charges (*two protons*) creates a repulsive force, and between two unlike charges (*electron and proton*) an attractive force.

It is difficult to fathom how the exchange of a particle of any kind can create both an attractive and repulsive force. The theory is that when an electron or quark emits a photon, it changes that particle's velocity, and when another particle absorbs the photon it changes that particle's velocity, *just as if* there had been a force between the two particles.[3] It is especially difficult to believe the exchange creates a consistent force when the exchange particle in question is a *virtual* particle. The following observations make the Standard Model's force theory by virtual particle exchange difficult to believe.

- A virtual particle is one that theoretically pops in and out of existence. How can there be a consistent and smooth *force* while relying on a particle that is there one moment and gone the next, no matter how quickly the exchanges take place? This is especially mystifying considering each charged particle is theoretically exchanging an untold number of photons with every other charged particle at the same time.

- *Virtual* particles are hypothetical, never directly observed. They are only believed to be real because there is no other reasonable explanation for such things as the Casmir Affect (*two metal plates being pushed together by unseen forces*)[4], and because mathematically if the virtual particle's presence is not included in the calculations the answers are wrong.[5] So, the force by virtual-particle exchange theory relies on a hypothetical particle having a hypothetical effect on another particle.

- Virtual-particle exchange often results in calculations that include *infinities* and the only way to make sense of the calculations is to simply ignore or remove them. Removing them has been given the innocuous name of *renormalization*; a procedure that is admittedly dubious mathematically.[6] None the less, the interaction between photons and matter is summarized in a body of equations known as Quantum Electrodynamics (QED), which has proven very accurate and is used throughout science and industry. Again, the Standard Model has found a way to make the math work, but that doesn't mean the underlying theory involving the exchange of virtual bosons is correct.

- Of the four virtual-exchange particles, the Graviton, theorized to convey the gravitational force, has never been detected. The absence of graviton brings doubt to the entire scheme of boson exchange being the source of force. Even if gravitons are miraculously found, the interaction between matter and gravitons produces infinities that cannot be renormalized, so at present according to Physicist Michio Kaku, graviton exchange is not even a viable theory.[7]

- Physicist Kenneth Ford and others tell us that every time an electron absorbs or emits a photon, entire new particles are created and the original particles are destroyed.[8] If the original emitting electron is turned back into a knot of Penergy and from that knot a whole new electron and photon are created, and this happens every time a photon is absorbed and emitted, it seems like a *very* laborious, inefficient, and unreasonable way for nature to work.

- The Tor Model posits force being conveyed through the fields of particles, and scientists have noted that fields take time to exert their influence. Physicist George Musser notes that a time lag seems odd if forces are leaping directly from one object to another [*as in particle exchange*] but a time lag is perfectly natural if an impulse must make its way through a medium.[9] In other words, one would expect a force conveyed through a boson exchange to be initiated smoothly, without a time delay, but evidently that is not what scientists observe. The noted time delay makes sense if forces are conveyed through fields in a medium.

- Two charged particles theoretically feel a force between them, even at galactic distances. This makes an electron on earth theoretically having a very small influence on an electron on Alpha Centauri, four light years away. If this is so, it would mean that one electron is constantly exchanging virtual photons with the other.[10] It would also mean the electron was making the same constant exchange of virtual photons with every other charged particle in the space in between! Again, that makes this scheme laborious and inefficient, making it difficult to believe nature works that way.

- Theoretically, for virtual particles to come into existence they borrow energy from what is called the vacuum of space. If virtual particles are popping in and out of existence constantly to affect forces, it would require a great deal of energy to bring all of them constantly in and out of existence, even momentarily. According to Tim James, quantum theory can predict how much energy density there should be in the vacuum of space by adding all the virtual particles together. Doing so produces a theoretical energy value of 10^{105} Joules (*think ounces of energy*) per cubic centimeter. Examining space, however, produces a value of only 10^{-15} Joules per cubic centimeter. Comparing what the energy level theoretically should be to what is measured, scientists find the theoretical value to be 10^{120} times larger than the actual value.[11] Michio Kaku characterizes this as the largest mismatch in the entire history of science.[12] Virtual particles might exist, but apparently not in the numbers to support them being the source of all the forces.

According to quantum physics, the vacuum of space is awash in virtual particles constantly popping in and out of existence.[13] The Tor Model recognizes the possible existence of virtual particles but not to the degree theorized by quantum physics. In the Tor Model, virtual particles only come into existence when a fluctuation in the Penergy medium creates a small knot that provides a momentary rise in the Penergy density sufficient to bring particles into being. But because the knot begins immediately dissipating back into the surrounding thin-density Penergy medium, the failing knot is not strong enough to sustain itself, so the particle(s) disappear back into the medium before being fully formed.

Consequently, virtual particles are not popping in and out of existence everywhere, only where fluctuations in the Penergy medium are prominent such as near blackholes and other very dense masses, or within the environment of a planet's atmosphere such as our own. Charged particles may absorb and emit photons, even virtual photons that affect each other's momentum, but that fact by itself does not necessarily create an electric *force*.

We will now examine force from the perspective of the Tor Model as being a relationship between personal fields, and we will look at how that works in each of the forces.

Force According to the Tor Model

In the last chapter we discussed the first significant relationship having been between particles and the thin Penergy medium that created particle *fields*. The next significant *relationship* to evolve in the universe was between the respective particle fields. The Tor Model sees the relationship between fields as the source of *force*. A force affects all particle interactions and allows particles to combine. To understand how fields produce a force, let's examine which force might have allowed Tors to combine and how that would have been accomplished.

The theoretical Strong and Weak forces only involve subatomic particles confined to the nucleus, and since Tors are present in all particles, those forces cannot be those that hold Tors together. The gravitational force is the attractive force between masses that holds us to the surface of the earth, but it is far too weak at the particle level to be a strong combining force.

The Electromagnetic force is a combination of the electric force that pushes electricity and the magnetic force that draws magnets together. Electric force involves the relationship between two different charges, positive and negative, requiring two different particles, but we are looking for a force that works on a single particle. The magnetic force might emanate from a single particle, so let's examine the magnetic force to see if it could be a candidate for the force first allowing particles to combine. In doing so we will explain how particle fields are the source of force.

The Magnetic Force

Scientists have observed that when magnetic attraction comes into play the particles involved are lined up in the same direction, meaning the axes of their spins are aligned.[14] A natural magnet called lodestone (*magnetite*) is *magnetized* by virtue of the electrons in its atoms having their spin axes aligned. When the lodestone is placed near a metal object, the magnetic field around the lodestone causes the spin axes of the metal object's electrons to align in the same parallel position.

The magnetic body with its electrons so aligned is said to have a head and a tail, or a north pole and a south pole. When the north pole of one magnet is placed in line with the south pole of another magnetized object, they are attracted to each other. But when they are aligned otherwise, especially north to north or south to south, they repulse each other, as you know if you've ever played with magnets. But what is causing this unseen force aligning the spin axes? Where is the force coming from?

Force was once believed to be an undefined inherent aspect of certain pieces of matter, namely metals, and has since been ascribed to boson exchange. The Tor Model sees the magnetic force as the relationship between the *fields* of like-spinning Tor particles. The Tor particle spin causes the surrounding Penergy to spin with it and spread out creating a field. When the spinning field of one Tor comes into a nose to tail alignment with the like-spinning field of another Tor particle the fields are attracted to each other and will combine due to their natural coupling capacity. If the spin alignment is otherwise, the fields repulse each other. This is the basis for the magnetic force, and it is due solely to the spin dynamics and fields of the respective like-spinning Tor particles.

The force is evident in our observations of a *magnetic field* represented by iron filings on a paper placed above a bar magnet, as shown in Figure 4.1. The magnetic field lines loop through the north and south poles, entering through the south pole and exiting through the north pole. When two bars are placed in line, those field lines continue through both bars creating a natural attraction between the north and south poles. We also observe that the strength of the field is greatest near the poles and falls off as one moves away from those poles.

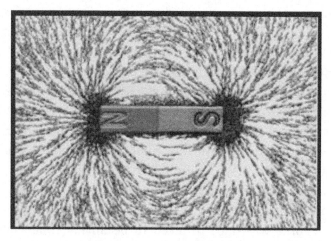

Figure 4.1 Iron filings on a sheet of paper covering a bar magnet showing its magnetic field.

This picture of iron filings is a good example of particles producing a field around themselves, and the cumulative field influencing other particles. In this case the aligned atoms within the magnet are producing a magnetic field that is in turn influencing the atoms in the iron filings, causing them to move into a fixed orientation consistent with the magnet's field.

In summary, the *fields* of two Tors will cause the Tors to be attracted to each other if their fields are aligned nose to tail and repulse each other if meeting in any other arrangement.

Creation of Tor-Chains

The magnetic force generated through Tor fields is what brings Tor particles together. When aligned nose to tail they are attracted with sufficient strength to hold each Tor in that attached relationship. Such in-line attachments would create a *chain* of like-spinning Tors, as illustrated in Figure 4.2.

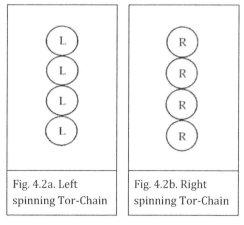

| Fig. 4.2a. Left spinning Tor-Chain | Fig. 4.2b. Right spinning Tor-Chain |

Figure 4.2 The strong nose-to-tail coupling between like-spinning Tor fields creates the magnetic force and allows Tors to combine into Tor-Chains.

Tor-Chains can be left-spinning or right-spinning and can be any number of Tor particles long. Longer chains would make it less likely the chain would be annihilated by bumping into an opposite spinning chain of the same length and spin signature. However, the longer the chain the more likely it would be broken apart by a collision with a more energetic Tor or Tor-Chain.

Deduction #9: Given the presence of personal particle fields; and given that Tor particles spin on their own axis at a very high rate creating a field with a strong coupling capacity; we can conclude that the source of the magnetic force is inherent in the Tor field. The fields of like-spinning Tors create a strong attractive force when Tors are aligned nose to tail, which allows for the creation of Tor-Chains.

A Tor-Chain could be hundreds, even thousands of Tors long. In the early universe these long chain creations probably happened, but the utility of such long chains is questionable. On the other hand, proteins are long chains of nucleic acids encompassing thousands of atoms that have proven to be very useful if not essential to evolutionary growth. Tor-Chains may turn out to be equally essential to evolutionary growth and we should be open as to how many Tors make up a viable/useful chain.

The Tor-Chain creates a surrounding field, the effect of which causes other nearby Tors to align their spin axes with that of the Tor-Chain. This spin-alignment force carries over to electrons and atoms, as well. Consequently, bound particles such as electrons can have their spin axes aligned, and the alignment of multiple-particle spin-axes creates a recognizable magnetic field. The field is attractive to another magnetic field when the axes are aligned nose to tail. Again, the more particles involved in the alignment process, the greater the size and strength of the magnetic field.

The Electric Force

Eighteenth century scientists and others including Ben Franklin were familiar with the rudiments of electricity being the motion of electric charges which Franklin labeled positive and negative. French scientist Charles-Augustin de Coulomb established in the late 1700's that charged objects attract and repel each other. The connection between electricity and magnetism was discovered in the early 1800s by Michael Faraday. Faraday formulated the notion of magnetic and electric fields, and that particles passing through those fields would experience a magnetic and electric *force* capable of changing that particle's velocity.

The source of the force was initially mysterious and was thought to be an inherent quality in the charged object. With the advent of quantum theory, the electric force was ascribed to the exchange of virtual photons. The Tor Model posits the electric force to be inherent in the relationship between Tor particles. In the Tor Model, the dynamics of the spinning fields cause opposite-spinning Tors to be attractive and like-spinning Tors to be repulsive, except when aligned nose to tail.

It is known that elementary particles with opposite spins are subject to an attractive/repulsive force; particles identified as matter and anti-matter are subject to an attractive/repulsive force; and particles said to possess opposite charges are subject to an attractive/repulsive force. In all these cases the source of the force is the same – the electric force created by the relationship between the fields of Tor particles.

In summary, what we know as positive verses negative charges, and matter verses anti-matter, is simply particles comprised of various Tors. The respective fields of the Tor particles create an attractive/repulsive *electric force,* with opposite spins being attractive and like spins being repulsive, except when Tors are aligned nose to tail.

Particle Assembly from Tor-Chains

Since mirror-image particles such as Tors have a natural attraction for each other, it is reasonable to conclude that mirror-image *Tor-Chains* would likewise have a natural attraction. Exact length mirror-image Tor-Chains would annihilate, but if they had different lengths, they would have different spin signatures and therefore would attract each other without annihilation. The attraction allows opposite-spinning Tor-Chains of different lengths to successfully combine, as illustrated in Figure 4.3.

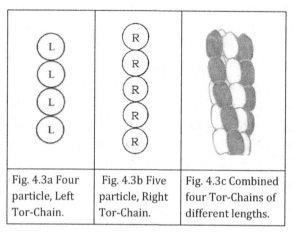

| Fig. 4.3a Four particle, Left Tor-Chain. | Fig. 4.3b Five particle, Right Tor-Chain. | Fig. 4.3c Combined four Tor-Chains of different lengths. |

Figure 4.3. Left and Right Tor-Chains of different lengths have a safe attraction for each other and combine without annihilation.

Tor-Chain attraction could create combinations of various chain lengths with any number of chains involved. It may have been such a combination of various chain lengths that created the first modern particle, the *photon*, whose configuration is much like that shown in 4.3c. The configuration of the photon will be discussed in detail in Chapter Six.

With like-spinning Tors attached nose to tail creating Tor-Chains and a *magnetic field*, and opposite spinning Tors attracted to each other creating an *electric field*, we have the forces necessary to construct composite particles.

Creation of Tryks

Individual mirror-image Tors will annihilate, but when bound up tightly in a group such as a chain, the individual Tors are no longer subject to annihilation. Only a mirror-image chain of the same length and spin signature can annihilate the chain. The environment of many different chain lengths would have slowed the existing annihilation rate. This would have allowed the chains to survive long enough to combine and become a part of a larger composite.

Having Tor-Chains and single Tors available to combine, the next stable combination might be a long Tor-Chain *encircled* by any number of single Tors. For simplicity that notion will be illustrated in Figure 4.4 as a four-Tor-Chain encircled by a single, opposite spinning Tor. All kinds of different combinations are of course possible, and this configuration is set forth only as a working example. We will call this individual-Tor and Tor-Chain combination a **Tryk**. The mirror-image form of the Tryk configuration would also evolve and be stable, and as usual if the two should meet they would annihilate each other.

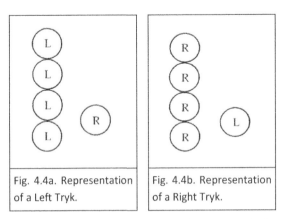

Fig. 4.4a. Representation of a Left Tryk.

Fig. 4.4b. Representation of a Right Tryk.

Figure 4.4. Representations of a Left and Right Tryk made from a Tor-Chain encircled by a single, opposite spinning Tor.

In a Tryk, the single opposite-spinning Tor is subject to the electric force and has the same encircling dynamics as that of an electron circling a nucleus to create an atom. Given that Tors can combine into chains, and opposite spinning Tors and Tor-Chains can attract each other, it is possible to have all sorts of combinations evolve if they are not mirror images of each other.

Recall that what we call charge is actually the dynamics between two or more Tor particle fields that give rise to the electric force. A *charged* composite particle, therefore, is comprised of a differential in the number of left and right spinning Tors. A Tryk with a dominance of left-spinning Tors (*shown in Fig. 4.4a*) therefore possesses a *positive charge,* and a dominance in right-spinning Tors (*shown in Fig. 4.4b*) possesses a *negative charge.* The strength of the electric charge is in proportion to the size of that Tor-spin count differential - the greater the Tor count differential, the greater the charge value. We will discuss charge values in detail in Chapter Six.

The Electromagnetic (EM) Force

Since Maxwell brought the magnetic and electric forces together under a single set of equations, the two forces have been thought to be two sides of the same coin. In the Tor Model they share the same repulsive force but differ in their attractive aspects. The magnetic force is attractive when like-spinning Tors align nose to tail, and the electric force is attractive between opposite spinning Tors. The electric and magnetic forces are distinct, but subatomic particles show evidence of possessing both. This makes sense, since all particles are made from combinations of Tor Chains and Tryks, which display magnetic and electric fields respectively, giving particles a magnetic and electric component. In those displaying an electric charge it is called the electromagnetic (EM) field/force. The EM force is normally measured by the strength of the electric field.

With respect to EM fields, their strength is proportional to the distance between the respective charged particles. The closer one brings the charges together, the stronger the force, but in the Tor Model this phenomenon has its limits. When the fields get too close, the fields interfere with each other and begin losing their attractive/repulsive capacity. If the particles were compressed to the point they were forced to touch, their fields would interfere to the extent they would presumably lose much if not all attractive/repulsive capacity.

Prediction: Someday an enterprising young physicist will create the experiment that will demonstrate that elementary particles when compressed together to the point of touching will lose much if not all their attractive/repulsive capacity, thus strongly suggesting that all attractive/repulsive force is derived from particle fields, not the exchange of virtual bosons.

Charged particles such as protons and electrons are naturally surrounded by their own field containing all their attributes, including the effects of their charge differential. If one lines up a row(s) of protons on one side and a row(s) of electrons opposite them, with their respective fields overlapping, it creates an attractive electric differential that will induce electrons to move across the combined fields creating electricity.

Given a reasonable distance between them, the more protons and electrons that are involved, the stronger the combined field. The strength of the electric differential between the positively and negatively charged rows is measured in *volts*. An electric charge differential can also be *induced* in a wire by the movement of an adjacent magnetic field, which is how a generator works.

The electromagnetic force is responsible for the attraction between electrons and protons, a combination necessary to create atoms, for atoms to create molecules, and for molecules to create life. The electromagnetic force is also what we see in action all around us daily. The repulsive aspect of the electromagnetic force is what prevents us from falling through the floor. The molecules in the floor are holding on tightly with the electromagnetic bonds that repel our shoes, holding us above the floor. Obviously, electromagnetic forces are very important in our daily lives.

The Force (F) between two charged particles falls off inversely to the square of the distance between the two charges (q). The equation expressing this notion is, $F=Kqq/\mathbf{r}^2$; (K being merely a constant). It's the $/r^2$ that represents the inverse square part of the equation, meaning at twice the distance, $/2^2$, the force is $1/4^{th}$ of what it was, and at 3 times the distance, $/3^2$, the force is $1/9^{th}$ what it was, etc. This point is important to remember as we turn to the discussion of the gravitational force.

The Gravitational Force

According to the Standard Model, *Gravitons* are the bosons mediating the gravitational force. That means that all the particles in your body and every other earthbound body are constantly exchanging gravitons with the earth. It would mean the earth is constantly exchanging gravitons with the moon and sun. It would mean the sun and every other star in the Milky Way is constantly exchanging gravitons with the SMBH at the heart of our galaxy. It's seemingly the only way all those gravitational forces could exist if gravity worked by exchanging gravitons. Wow, that is a lot of long-distance, continuous particle-exchanges, making it difficult to believe nature works that way.

The graviton has never been detected but not to worry, Einstein gave us a much more plausible explanation for gravity. Einstein showed us that gravity arises out of a curvature of the fluid-like nature of space, which bends the pathway of nearby matter relative to the size of the respective masses. Einstein's General Relativity tells us how gravity works with respect to mass, but that does not give us a clear explanation of what gravity *is*. Let's take a closer look at the possible source of the gravitational field.

The Gravitational Field

From Chapter One we learned that the only two distinct stable things that exist in the universe are blackholes and Tor particles, both of which are said to possess mass. We know that gravity acts like an attractive force due to the curving of the fabric of space (*the Penergy medium*) in proportion to mass. The Penergy medium is geometrically flat except where a gravitational influence causes a curvature, and Einstein gave us the mathematics to calculate how much distortion is taking place and to what degree that distortion will affect a change in the velocity of passing masses. But how does sheer mass distort and curve space? What is inherent in mass alone that could possibly create a gravitational field? Perhaps, it is not the mass itself that is curving space, but something else that exists *in proportion* to mass.

We know that massive, spinning objects such as blackholes curve space.[15] This observation could be important. It might be the *spin* of the object that is pulling on and curving the Penergy medium, not simply the object's sheer mass. That reasoning may work for blackholes that are known to possess significant spin and are known to drag the surrounding space with that spin, but how would that work for tiny particles that are only part of a massive object, especially one that may be spinning relatively slowly such as a planet? Where would its gravitational field come from?

Mass is simply an accumulation of particles, and those particles are all comprised of combinations of Tors. The only way any mass has contact with the Penergy medium is through the surface of those spinning Tors. Tors spin on their own axes and presumably at a very high rate. Tor spin therefore might be capable of tugging on the Penergy medium to curve it, which we recognize as a gravitational field. Perhaps gravity on an object of mass is no more than the cumulative effect of all the Tor spins simultaneously tugging on the Penergy medium.

This is counter intuitive and there is no direct evidence for it, but there are indications that suggest a connection between spin and gravity.

- There are two types of mass. Mass that is measured by its resistance to movement is called *inertial mass*, and mass measured by its gravitational field strength is called *gravitational mass*. These two types of masses are based on different concepts, but they always measure exactly the same. Astrophysicist Stuart Clark tells us, "There is no reason why these two masses must be the same... This is the most mystifying thing in the whole universe."[16] As we will discuss in Chapter Five, the source of inertial mass may well be due to the gyroscopic effect created by the very-fast spinning Tors. As described above, the source of gravitational mass could also be due to the spinning of Tors. Tor spin could very well be the nexus between mass and gravity and be the reason inertial mass and gravitational mass have the same value.

- Another observation associating gravity with spin has to do with how star formation is initiated, a process called *accretion*. It is theorized that a large body of cosmic gas with enough density will begin to swirl on its own and as it swirls it will draw in more of the surrounding gas. This happens naturally and seemingly the swirls begin taking in more gas before a significant central mass can form. So, what initiates the process of accretion? And how does it draw in surrounding gases before establishing a central mass that could support a gravitational field? Experiments reveal that rotating objects display an inward acceleration call *centripetal force*. The cause of the centripetal force is customarily ascribed to gravity, but in the case of accretion it appears gravity *is* the centripetal force - initiated solely by spin.

- Further evidence associating spin and gravity is found in the fact that the strength of the gravitational force falls off at exactly the same rate as the electromagnetic force: inversely to the square of the distance between two masses (m).[17] The equation expressing this fact is $F=Gmm/r^2$; (*G being merely a constant*). The similarity in force dissipation between these two unrelated forces must be more than a coincidence. Two distinctly different forces that lose strength at the exact same rate suggests they have some underlying feature in common. Recall that the attraction/repulsion of the electric force is due to the dynamics of Tor spin. Now we are saying the nexus between mass and gravity could also be due to Tor spin. Tor spin being the source of both the electric force and gravitational force could be the reason the dissipation of the field strength of the electric and gravitational fields are identical. Both are derived ultimately from the same source - Tor spin.

The attractive force of electric charge occurs *directly* between Left-Tors and Right-Tors, each directly sensing each other's field. However, the attractive force of gravity occurs between the spin of Tors and their cumulative effect of curving the surrounding Penergy medium, which *in turn* affects other particles. This *indirect* effect of gravity on other particles could be the reason why it is much weaker than the other forces.

Deduction#10: From the observational evidence stated above we can conclude that gravity is a manifestation of the curvature of spacetime, and that the spin of both Tors and blackholes is responsible for spacetime curvature.

Prediction: Once the presence of Penergy, Tors, and personal fields is established, a sharp young physicist or astrophysicist will work out the mathematics to support, and the experiment to prove, that a gravitational field, like all other fields, is created by the spin of either particles or blackholes.

In summary, all fields and forces are a result of spin or the relationship between spinning bodies. The magnetic, electric, and gravitational fields and their related forces, as well as centripetal and centrifugal forces, are all derived from something spinning.

Scientists have noted unusual galaxy rotation speeds and have theorized that the additional mass responsible is coming from the presence of *dark matter*. Some of that added gravitational influence, however, could be coming directly from the supermassive blackholes at the heart of the galaxies. In the Tor Model, their gravitational strength can be greater than their mass if the denser Penergy fuel inside of them drives their spin rate up disproportionately creating their own *dynamic gravity*. Dark matter and dynamic gravity will be discussed in Chapter Seven.

The Strong Force

The Strong Force According the Standard Model

According to the Standard Model, the three quarks of a nucleon are held together by the *strong force*, which is generated by the constant absorption/emission of virtual gluons. Gluons have not been seen directly but are believed to have come into existence momentarily in the Large Hadron Collider based on the analysis of high-energy particle collision debris. The strong force is unique in that the three quarks cannot be separated no matter how much energy is put into the effort. It is as if they are held together by strong rubber bands.

Quarks and gluons are theorized to possess a special kind of charge called Color Charge, which controls how quarks exchange gluons. A color charge has never been detected and is derived solely from a brilliant scheme developed by physicist Murray Gell-Mann in the 1970's that explains very accurately how this force might work. The Strong Force operates at very short distances and only inside the nucleus.

The theory and math supporting it are quite complex, there being three different color charges and eight different gluons involved. A good summary of quarks and gluons and how they were theorized can be found in Timothy Paul Smith's, *Hidden Worlds – Hunting for Quarks in Ordinary Matter*.

One of the ways particles combine is by sharing a part of themselves, like atoms sharing an electron to create molecules. One could argue that quarks are sharing a part of themselves, a gluon, but it's not quite the same. When atoms share electrons they create a momentary charge imbalance, and it is the charge imbalance that is holding the atoms together, not the mere sharing of the electron. In the case of the gluon exchange, because gluons are theoretically electrically neutral, sharing a gluon would not create an electric charge imbalance.

The Standard Model overcomes this by theorizing the Color Charge and the need for the composite proton or neutron to be 'color neutral'. That means a composite particle having three quarks must have one and only one of each of three colors per quark. A composite particle comprised of two quarks must have a color and an anti-color in the two quarks. The source of color charge is unknown.

The Strong Force According to Tor Model

It remains difficult to believe there is a force that holds two or more particles together by exchanging virtual particles. There is also the argument that every time there is an absorption or emission, all particles involved are destroyed and new particles produced, creating the height of inefficiency. Consequently, it is worth exploring an alternative theory for the source of the strong force.

Scientists now believe a nucleon contains far more matter particles than just the three quarks defining its electric charge value. This makes sense because the mass of a nucleon is perhaps a hundred times the mass of the three defining quarks combined, so it is reasonable to surmise that there is more to the nucleon than just three quarks.

To investigate the interior of the proton, scientists shoot high energy particles at it and observe how they scatter off the proton's interior. The more energy the shooting particle possesses, the smaller its wavelength and the better definition the scattering theoretically provides.

In a 2021 Forbes Magazine article by Astrophysicist Ethan Siegel entitled *What Rules the Proton: Quarks or Gluons?* he details what scientists find from those scatterings. He reports that low energy collisions are dominated by quark-quark interactions. Higher energy collisions start to see quark-gluon interactions, with some quarks turning into heavier charm quarks. Still higher energies are dominated by gluon-gluon interactions. He summarizes this by noting that at low energies a proton is more 'quarky', but at higher energies it's rather 'gluey'.

This method of experimentally exploring the interior of a proton raises concerns. Perhaps shooting high energy particles into a proton doesn't necessarily give us a better definition of its innards but reveals the type of particles that could be *created* from the knots of Penergy produced by those high energy collisions. If the 'gluey' gluon state only shows up under high energy conditions, it suggests that gluons are not fundamental but only exist when there are knots of high density Penergy to produce them. The high energy collisions that create knots of Penergy are then decaying into smaller particles producing jets and the outline of particles that could be interpreted as being specific particles present inside the proton. These experiments might be telling us exactly what the scientists say they do but given how easily knots of Penergy can create particles we must be careful in our analysis of such experiments. Shooting high energy photons into a proton could be *producing* particles rather than merely identifying existing particles.

The only force the Tor Model recognizes acting like a rubber band is the very-strong magnetic-force working at the Tor level. As observed earlier, the creation of Tor-Chains is due to the powerful coupling capacity of the aligned fields of Tors spinning on their own axes. Such a force might act like a rubber band if one tried to pull a Tor-Chain apart. This would be especially true if the connections were stranded into multiple Tor-Chains, as illustrated in Figure 4.3c.

This notion makes sense once we examine the possible spin dynamics of the Tors. Imagine the spinning Tors affecting the surrounding Penergy medium by causing it to rapidly swirl in the direction of the Tor spin. As the nose of one Tor field came close to the tail of another Tor field, the swirls would align themselves and the natural coupling attraction would pull them together, uniting the two particles through their fields.

The two fields would pull into each other without the two particles ever touching. As the particles approached each other the strength of the attractive coupling would relax and perhaps even become repulsive if the two particles moved too close to each other. The two particles would remain in this relaxed, balanced state until some outside force affected them. If an outside force tried to pull them apart, the strong coupling capacity of the fields would again come into play resisting any separation as they were drawn apart. A force sufficient to overcome the strong coupling capacity could separate the two fields, but as observed in our particle accelerators, when enough energy is put into the effort to separate the quarks, the knot of Penergy creates new particles that fill in the gap before the quarks can be separated.

This scenario is a good argument for the quarks being connected by Tor Chains, however, a complete theory may be better developed once the additional internal parts of the nucleon, if there are any, have been better defined. We must await further experimental developments before a complete theory can be developed.

The Weak Force

A stand-alone neutron is unstable and will decay into a proton, electron, and neutrino within a few minutes. Scientists were familiar with the magnetic, electric, and gravitational forces, so to explain how a neutron could decay into a proton, scientists had to define one more force.[18] A new particle had to be introduced to explain this force.

Large particles were predicted and ultimately detected in the Large Hadron Collider, dubbed the W and Z particles. They have not been observed directly but are believed to have been identified by reading high-energy particle collision debris.[19] According to the Standard Model these W and Z virtual particles are the force carriers of the Weak Force that is responsible for the decay of the neutron and all other particle interactions inside the nucleus.

The decay of a neutron into a proton is actually accomplished by a down quark inside the neutron decaying into an up quark. In the Standard Model quarks are elementary particles, yet the down quark is known to decay into an up quark, electron, and neutrino. The only way for a neutron to theoretically decay into smaller particles would be for it to first change into a W- particle that would then decay into the smaller particles.

The W and Z particles are large, being more than 80,000 MeV, compared to the light-weight quark at less than 5 MeV. The W and Z particles exist only briefly before decaying. They are only seen in the Large Hadron Collider under very high energy conditions, yet they are theorized to come into existence every time there is a decay/creation or absorption/emission within the nucleus in moderate energy conditions.

Physicists have created the mathematics to support the notion these very massive particles come into existence by momentarily *borrowing* the requisite energy from the *vacuum of space*. As noted in our discussion earlier, apparently the available energy for all those borrowings is not actually present in the vacuum of space.

These massive particles may momentarily exist in colliders, but they might simply be first or second-generation creations that could not continue to exist in the thinning Penergy density as described in Chapter One. They may have gone the way of all other first and second-generation particles, making their relationship to today's particles questionable.

A more reasonable approach to down quark decay is for the parent particle to be comprised of the residual particles. Since composite quarks of the Tor Model do not require a special particle to decay, and the Tor Model's belief that down quark decay may be triggered by the absorption of a high energy photon, there appears to be no value in further guess-timating the W and Z particle's makeup or purpose.

There is also a *Nuclear Force* that holds the nucleus together, but that will be discussed when looking at the configuration of a quark in Chapter Six.

* * *

Using the electric force as an example, we examined the feasibility of force being effectuated through the exchange of virtual bosons and concluded that the observational evidence and related rational conclusions bring the theory into serious doubt. We next examined the Tor Model's theory for force being generated through fields created by particles and we examined each field to see how that would work. From these examinations, we concluded that the attractive/repulsive characteristics involving matter/anti-matter, electric charge, and magnets are all derived from the same source – the attractive/repulsive relationship between the underlying Tor fields.

We next examined the gravitational force and expressed an alternative theory for the source of the gravitational field being the cumulative effect of the spin of Tors making up a mass. We will put off our examination of blackhole mass until Chapter Seven as part of our discussion on *Dynamic Gravity*. We next examined the strong force and speculated on the source holding quarks together being the strong coupling capacity of Tor chains. Finally, we concluded that since quarks are not elementary but composite particles, there is no longer a reason to speculate on the existence of another force being necessary to explain the decay of quarks and other particles.

We have completed our discussion of the basic elements comprising the very earliest part of the universe's evolution. Our story began with only two initial conditions: *it started as a very condensed speck of pure energy*, and *the speck of energy began expanding*. Without the need for anything popping into existence or requiring mathematical explanation, we created a Tor Model that reasonably explains the *natural evolution* and origins of SMBHs, Early Elementary Particles, Dark Energy, the Cosmic Web, Fields, and Forces. In sequence, each of those creations *naturally evolved* from the conditions that evolved before it.

Theories being what they are, this sequence of natural evolution is perhaps no more valid than the Standard Model's Hot Big Bang Theory. But the Tor Model, despite its several *speculations*, seems to solve a few mysteries, is more elegant due to its simplicity and naturalness, and is more believable since it requires nothing magically popping into existence or other ad hoc infusions. The model introduces nothing that was not already known to science, but of course some aspects of the model have been given a new significance.

A good theory will require math to back it up. For now, we are simply investigating the theory to see if the time investment creating the math is justified. The hypotheses supporting the Origins we have developed thus far are consistent mathematically with much of the Standard Model's math. Although the source of fields and forces may be different, the mathematics supporting the actual interactions between particles should be very consistent. So far, the four recognized points of observational evidence supporting the Big Bang Theory are not inconsistent with the evidence supporting the Tor Model. This is because the events of the Tor Model took place before the creation of atoms, which is the starting point for the observational evidence supporting the big bang.

The next major development in the universe's evolutionary sequence is the Origin of Composite Particles, which naturally evolves from the creation of the fields and forces. Our discussion will include a new theory for *cosmic homogeneity*, the origin of *natural selection*, and the origin of *mass*. We will then discuss the possible configurations of subatomic particles, and finally the origin of the atom.

Evolution is not simply the story of how things changed but why they changed. Understanding nature at that depth gives us a glimpse into what is driving cosmic evolution and what that evolution in the future might entail. Be prepared for more new ideas, a new model for elementary particles, a new definition of cosmic evolution, and the resolution of other cosmic mysteries. The image of our new vision of the early universe is about to become more vivid.

Chapter Five
The Origin of Composite Particles

We are entering the perilous arena of hypothesizing new particles without a means to prove their existence. The motive driving this hazardous venture is that doing so provides answers to several outstanding mysteries and it completes the vision of how our Sub-A's *evolved* rather than having inexplicably popped into existence.

Current particle theory denies the existence of particles within our Sub-A's, but many Physicists don't reject the possibility. Physicist Jon Butterworth points out that what we think of as fundamental particles might simply be the smallest things we can measure with the tools available, and there may be deeper layers.[1] Physicist Bruce Schumm tells us, "No one claims that quarks have been shown definitively to be the fundamental building blocks; it's just that with any experiment that we can do today, we can't "see" anything smaller than roughly 10^{-18} meters across, and quarks are apparently smaller than that."[2]

The idea of possible composite Sub-A's was momentarily popular in the 1970's. Then the elemental particles were called *Preons*. Several of the Preon theories were believed to be quite elegant and provided answers to outstanding questions within the Standard Model. Those theories, however, could not explain the unknown force holding the particles together and the theories lost their following by the 1980's.[3]

The Tor Model has no problem explaining the force holding elementals together. Notwithstanding the risks venturing forward, the idea of composite Sub-A's answers too many questions to be ignored, so let's go where no one has successfully gone before and pursue the composite particle hypothesis and see where it takes us.

The evolution of our early universe has been quite a journey already. Our reconstruction project has undertaken several evolutionary steps, and we are still working on the building blocks of the first level – the Tor Level. Our universe is now filled with developing supermassive blackholes, cascade remnants of faltering mid-size swirls, Tor particles, and a very thin density of Penergy making up our interspatial Penergy medium. The Tor fields, endowed with what we now call the magnetic force, allowed Tors to combine nose to tail to create Tor-Chains. Through the electric force, Tor-chains could then combine with a single Tor to create a composite particle we are calling a Tryk.

The continuous creation and destruction of many Tors, Chains, and Tryk combinations would constitute the evolutionary phase of the next building blocks. Many combinations would come into existence, be broken apart, recombine into different configurations, and be broken apart again. It's the same product strengthening process every evolutionary Level goes through before a stable, dominant form eventually evolves from the frenetic environment. This evolutionary process establishing which combinations survive and go on and which do not is called *natural selection*. It is one of many processes taking place in the cosmic environment.

Before exploring how composite particles may have evolved, let's take a closer look at the cosmic environment, the stage upon which elementary particles will come together to create our Sub-A's.

The Cosmic Environment

Survival by Natural Selection

The actual combination of Tors and Tor-Chains in the configuration of Tryks is unknown. They may be quite complex. Those illustrated in Figure 4.4 are simply representations put forth as a working model. The dominant Tryk configuration will ultimately depend on its ability to survive within the cosmic environment. With all the creation, annihilation, and combining/destroying of particle combinations taking place, the thin Penergy medium would have heated up, becoming a hot, high-energy *frenzy* of activity. Only the strongest particle combinations would survive the frenzy. In this early stage of cosmic evolution, survival by *natural selection* had begun.

Biological natural selection is the process of organisms undergoing physical change due to DNA mutations. Some mutations make the organism better adapted to its environment, which raises its chances to survive and reproduce, perpetuating those survival traits. In the case of elementary particles, the strongest and most suitable composite combinations would survive the frenzied environment and go on to become stable and available to combine again, perpetuating its design. Collisions breaking up weaker connections that reconfigure themselves is the mutation equivalent for particles.

The third Level of matter will emerge from the frenzy of the environment to be quite complex compared to the first two Levels. For the third Level to evolve, many kinds of Tryk combinations will be made and torn apart, with their fragments recombining to form yet other combinations. Our Penergy medium will have become a sea of Tors, Chains, Tryks, and their combinations, much like the earth's oceans were a sea of macromolecules prior to cells evolving. All this particle creation, annihilation, and smashing combinations apart creates a Penergy environment that is truly a *frenzy*. It's a wonder any combination of Tryks makes it, but given enough time and testing by natural selection, viable combinations will prove stable enough to survive. Which Tors, Chains, and Tryk combinations would survive annihilation or be broken apart by collision is impossible to say, but natural selection in this hot, frenzied environment would definitely determine their number and composition.

Cosmic Homogeneity

By this time in our cosmic evolutionary history the environment had become a hot busy place with lots of diverse activities. We next address a cosmic environmental issue that has plagued cosmology for a long time: Why does the universe today appear so homogenous? Homogenous means something having its components (*like stars and galaxies*) uniformly distributed throughout.

According to Astronomer Carolyn Devereux, that question is known as the cosmic *horizon problem* (aka *homogeneity problem*) and is one of the most important problems in cosmology.[4] The leading theory to explain the universe's homogeneity has not been made a part of the Standard Model of Cosmology but is still widely popular. The theory is rather incredible and deserves our examination. The homogeneity issue also addresses the sequence in which the universe was built, so it is important for our reconstruction project to get it correct. Let's examine homogeneity as expressed in the leading theory and how it would look if the universe were built from the Tor Model's blueprint we've developed so far.

Cosmologists presume the universe is uniform and appears the same from any vantage point within the universe.

Over a distance of 300 million light years, from any point in the universe the galaxies and distances between them look much the same, the stars and their contents look much the same, etc. This presumption is so strong it has been named the *Cosmological Principle*. This uniformity, however, is not consistent with the universe envisioned by the Cosmological Model, which does not predict a uniform universe, making the horizon problem a fundamental issue. According to the Standard Model, the star clad galaxies did not start to form until roughly a billion years after the big bang.[5] By this time the universe was too big for it to balance out and have its contents uniformly distributed to the degree we see today.

Imagine a large, cold room with a thermostatically controlled wall heater. The temperature drops, the heater goes on and begins to heat the room, creating hot and cold areas as the heat disperses. Given enough time, the temperature in the room would equalize and be fairly uniform throughout. But what if the room was constantly expanding at a rate faster than the equalizing effect could take place. The temperature could never reach uniformity. This is the scenario scientists have calculated for the early universe. Under the Standard Model's scenario, no matter how far one goes back in time, the universe was never compact enough for a long enough time to achieve homogeneity.[6]

To overcome this and other conceptual problems within the Standard Model, Physicist Alan Guth in the 1980's put forth a radical explanation called *inflation*. He theorized that within two hundredths of the first second of the big bang the universe blew up to a comparatively enormous size *very* rapidly. Stephen Hawking described it as if a one-centimeter-wide coin suddenly blew up to ten million times the width of our Milky Way galaxy.[7]

We don't know what started the rapid expansion or what stopped it. The expansion evidently solves the homogeneity problem but stretches believability. The theory is complicated, involves undetectable fields, the creation of new, undetected particles (*inflatons*), and has other ad hoc characteristics.[8] This is another theory that may work mathematically but is difficult to believe. Tim James in his book, *Astronomical,* points out, 'Inflation is by no means an accepted theory and it generates all sorts of unanswered questions of its own...'.[9] Physicist Sean Carroll reminds us that there is no evidence that *inflation* ever happened.[10]

Let's explore an alternative scenario for the universe's homogeneity by looking again at the Cascade Effect, and SMBH and particle formation. The process begins with the universe consisting of many proto SMBHs swirling into existence with marginal gravitational influence. As detailed in Chapter One, at the edges of the giant swirls, more swirls start their motion, and at their edges more swirls begin, cascading into smaller and smaller swirls, ending in composite particle generations at the guesstimated .15, .10, and .05 Penergy densities.

All this activity took place universe-wide at the edges of the cascade of proto-blackholes, which did not yet exhibit sufficient gravitational influence to draw inside much of the newly created particle matter. The newly minted early elementary particles were free to roam the entire universe and begin combining. Once the bigger swirls gained sufficient angular momentum and a serious gravitational influence, the free particles were drawn closer. As the swirls developed into blackholes, their spin caused the surrounding particles to swirl, flatten to a disc, and form the structure of our galaxies.

All of this suggests that SMBHs, particles, and perhaps early stars, completed formation at roughly the same time. Most of the mass of a galaxy is in the content of the surrounding particles and stars. The SMBH generally represents only 1% of the galaxy's total mass. Apparently, the SMBHs drew inside very little of the early mass of particles but drew around them exactly the right amount to support their gravitational field as noted in Chapter One.

This scenario seems consistent with findings of scientists looking at some of the oldest galaxies that date back twelve billion years. As reported in an article dated 1/27/2021 in *Scientific American* entitled *Giant Galaxies from Universe's Childhood Challenge Cosmic Origin Stories,* some of those very old galaxies are too large to have formed in the customary way of being built up from hydrogen gas and star debris. This article is consistent with the one discussed in Chapter One. Both articles say that scientists are challenged to find a reasonable explanation of how those galaxies could have become so big, so soon after the big bang.

To explain both the size and age of the galaxies only requires the proper timing for particles and particle combinations to come into existence just as the pure energy blackholes were gaining gravitational influence. By this time the universe was uniform in its contents and remained homogeneous right up until the time the Standard Model sees atoms created and photons decoupling from matter creating the Cosmic Microwave Background.

The gradual disappearance of the intermediate sized swirls and the gradual growth and development of the larger swirls might account for the miniscule ripples in the Cosmic Microwave Background, discussed further in Chapter Six. This description of SMBH development, particle creation, and the homogeneity of the universe all takes place without the need for the universe to undergo anything as radical as *inflation*. The early expansion of the universe can better be described as taking place during the time the SMBHs, cosmic web, and early elementary particles were evolving.

Because the SMBH creation sequence also easily answers both the homogeneity problem and how the cosmic web was formed, it again validates our first tentative deductions and allows us to make Deduction#11. It also emphasizes the importance of our understanding the dynamics of Penergy and Penergy density in our universe.

Deduction#11: The creation of Pure Energy Blackholes and the Cascade Effect appear to account for both the formation of the cosmic web and the homogeneity of the universe.

Next, we need to fortify our reconstruction blueprint with a clearer image of what is happening in the big picture of this project. What is shaping our construction? In other words, what is driving this cosmic evolution?

Evolution Progresses in Levels

There are many theories about evolution, each having some validity depending on their perspective. One such perspective is to think beyond planet earth; to step back and observe how matter has evolved cosmically over the entire history of the universe.

Going back to the evolutionary level we recognize as **Sub-A's**, [*first minute of the big bang*] scientists have theorized that in time the existing positively charged protons and neutrally charged neutrons *combined* to form nuclei. The nuclei eventually *combined* with various configurations of negatively charged electrons creating the next evolutionary level of matter – a variety of **Atoms**.

Once the atomic level stabilized, the atoms went through their own evolutionary stages, *combining* to form hydrogen gas molecules that in turn *combined* to form stars, and later stars and planets. On at least one planet (*earth*) molecules of atoms *combined* to form large macromolecules, creating proteins, carbohydrates, nucleic acids, and lipids, that eventually *combined* to form the next evolutionary level of matter – the **Cell**.

Cells in turn went through their own evolutionary stages developing sensory apparatus, locomotion, a circulatory system, a nervous system, and the ability to *combine*. These advancements lead to the creation of the next evolutionary level of matter – the **Organism**.

Based on observations of just these four evolutionary levels we might conclude that evolution progresses in stages to *Levels*, and that each Level is the result of the previous Level growing in complexity and *combining*. We could call this process the Combination and Growth Process. But since it has existed for billions of years and seems very precise in both its persistence and direction, if we can find evidence that matter has *evolved* in this way since the birth of the universe, we could more precisely call it the *Combination and Growth Imperative*, or the CAGI for short. We would only need to show that matter has *evolved* by this process, as opposed to simply popping into existence.

The constant building of complexity has long been observed by others. Physicist Paul Davies tells us, 'The fact that nature has *creative power,* and is able to produce a progressively richer variety of complex forms and structures, challenges the very foundation of contemporary science.'[11] His book, *The Cosmic Blueprint*, was published in 1989, but the notion is still relevant today.

There is plenty of evidence for the building of complexity, but there is also a force tearing things down. Some believe the force tearing things down is the dominant force and will be the one most influential in determining the destiny of the universe. Why would the universe build into its fabric a force to tear things down?

Entropy – Cosmic Recycling

Thermodynamics is the study of heat and energy. The reader is probably familiar with the first law of thermodynamics: energy cannot be created or destroyed. The second law of thermodynamics encompasses the idea that things and systems are usually wearing down, becoming less energetic and more disordered and random. The term to describe this phenomenon is *entropy*. There are many interpretations of entropy and what follows is only one of many perspectives on the subject.

Disorder implies a prior state of order. Throughout the universe both the building of order and disorder are taking place simultaneously, and for good reason. The Combination and Growth Imperative (*CAGI*) progresses in evolutionary stages to produce Levels of matter that are put together (*combined*) from the elements in the Level that preceded it. In this way, matter is constructed from building blocks that are constantly changing in size and complexity.

A growth in complexity requires structures to be configured in new ways. Their building blocks must be readily available and capable of combining with others to test which combinations are best suited for the environment in which they reside. The best building blocks are of course those that have already endured the tests of a punishing, hazardous, or frenzied environment, and are being recycled, fragments tossed into the boneyards of physical debris and the Penergy medium to be reused when needed.

Entropy is the universe's way of breaking down complex units into smaller, less energetic building blocks, making them available for reuse. Even the highly complex levels created by the CAGI process are subject to entropy. If nothing broke down at each level of complexity, we would soon run out of choice building blocks of matter and energy. The environment is constantly but subtly changing, requiring constant entropy to produce building blocks with the latest core, survival attributes. For complexity to grow effectively, death and decay must be a part of the process. No level or their building blocks are permanent. They are used by the CAGI process temporarily to further its progress and then recycled. It's just how evolution and the universe work. *Why* it may work that way will be discussed in the final chapter.

Cosmic Particle Attributes

Particle Emergent Qualities

To examine whether our CAGI theory has merit we must determine whether it is feasible for the early elementary particles to have combined in order to get that combination and growth process started. But Tors are simple, spinning tornados of Penergy; how can they possibly grow into the complexity of matter we see today?

Looking at a lone Tor particle, one could not visualize how a small, spinning tornado of Penergy could ever grow into something more complex. We will likely ask the same 'how could this possibly happen?' question at the start of each evolutionary Level because the dominant structures at the start of each Level always seem so simple, they show no evidence or promise of combining potential. What we later observe, however, is that each Level evolves new and exciting *emergent qualities* that assist it in changing, growing, and combining.

Emergent qualities come about according to Physicist Stephon Alexander when elementary constituents interact to create novel properties that are not possessed by the constituents themselves.[12] Looking at two hydrogen atoms and one oxygen atom, one could not anticipate their combining to produce a substance such as water. Nor could one anticipate the creation of all the other gases, liquids, and metals that can be produced from other combinations of simple atoms. Emergent qualities are at the heart of the evolutionary process.

The emergent quality in Tors was, of course, their effect on the Penergy medium that created *fields* giving particles the capacity to interact. Emergent qualities are important tools used by the CAGI in the advancement of cosmic evolution. We will identify additional emergent qualities as we build evolutionary Levels.

Particles Combining

In the model we are putting together, Tryks are the stable, second level of matter. To reach the third level, Tryks will have to combine to create a more complex form of matter. In the Tor Model, particles combine in two ways. Firstly, due to a natural attraction inherent in their fields such as the way like-spinning Tors combine nose to tail to create a chain, or the way the natural attraction between electrons and protons creates atoms. Secondly, two or more particles can also combine by sharing parts of themselves such as protons and neutrons sharing Pions and Gluons to create a nucleus, or atoms sharing one or more electrons to create a molecule. In our examples, Tryks combine using natural attraction.

Either of these methods will hold two or more particles together allowing for the creation of larger particles but other scenarios may be possible. According to the Standard Model, protons are made up of three quarks that are held together by a strong force that involves both sharing and natural attraction. Such a force may represent an entirely different way of combining, but its existence is only theoretical and not completely understood. In the Tor Model, all forms of combining are ultimately due to interactions between particle fields.

Particle Levels and Size

Tors represent the most elementary level of matter, so they are referred to as Level 1 particles. Tryks (*or something like them*) are built from Tors and represent Level 2 particles.

Sub-A's are built from Tryks and represent the Level 3 particles. There might be an intervening Level, but for simplicity let's continue with what we have and build our Sub-A's from the Tryks illustrated in Figure 4.4. Other scenarios are of course possible.

Tryks are very tiny, tightly wound particles. We may never see them or be able to prove their existence. Composite particles hold themselves together by a force through their fields, the strength of which is called their *binding energy*. It takes at least the amount of the binding energy to pull apart the constituent elements of a composite particle. Tors spin on their own axes and create a very strong field. Due to the strength of the Tor fields making up a Tor-Chain, the binding energy of Tor-Chains and Tryk-Chains is very high. We may never build a particle accelerator with sufficient energy to overcome the binding energy to break matter down to the Tryk level.

Tryks can combine to form particle chains of any length. Still quite small, these chains would look like minute bits of wiggling string. The wiggling, spinning, vibrating movements of each particle affect the Penergy around it, creating a field. These are all important attributes of early elementary particles, characteristics that will be carried forward, expanded upon, and prove essential to future Levels of evolution.

Particle Consistency

As far as scientists can tell, all individual electrons and protons are consistent in their mass and electric charge value. Those particles created anew out of knots of Penergy inside colliders have the exact same mass and electric charge value as their counterparts that were created billions of years ago. What can account for such consistency?

Tors were created within a *band* of Penergy having a guesstimated density of roughly .15, producing Tors with slightly varied masses. Only Tors of the exact same mass can combine (*or annihilate*), so once an optimal size Tor was eventually *naturally selected* to be dominant it would create consistent Tor-Chains and Tryks.

The other Tors and Tor combinations would eventually be annihilated, torn apart, or become part of the fragments of particles making up *dark matter*, which will be discussed in Chapter Seven. One naturally selected Tor size assured that subsequent particle types created from Tors and Tryks would have the exact same mass, charge, and other overall characteristics. Ultimately, particle consistency is a product of natural selection.

Particle Angular Momentum

Angular momentum is the rotational analog of linear momentum – an attribute of a body with both mass and velocity. Particles spin and therefore possess some measure of angular momentum. Quantum Theory's fundamental particles do not spin on their own axis. Their spin cannot be measured directly but is measured indirectly by a process using *projections*. QT particles are therefore said to have an *intrinsic amount* of angular momentum. Physicist Bruce Schumm tells us, 'Describing spin as "intrinsic" doesn't really tell us what it is'. In Quantum Theory, the physical aspects of particle spin are somewhat of a mystery. [13]

In the Tor Model, the only particle spinning on its own axis is the Tor. The model sees electrons and quarks not as fundamental particles but as composites. As composites, comprised of smaller particles in various spin orientations, electrons and quarks would not spin on their own axes but instead spin on a center-of-mass axis. Not having a true spin axis, it is difficult to measure any kind of true angular momentum. The absence of a true measure of angular momentum for electrons and quarks is better understood if they are recognized as composites rather than elementals.

In Quantum Theory there is a mystery as to why it takes a fundamental particle two complete revolutions to return to their original position. Composite particles may be the answer. Scientists have noted that spinning gyros in multiple orientations are stable if their moments of inertia are the same. If different, however, one of the orientations becomes unstable allowing the system to tumble in the unstable direction. Given the multiple spin orientations within a composite particle comprised of many Tor-Chains, instability in one of its orientations is inevitable, causing the particle to not only spin but tumble. If it tumbled one-half turn for every complete spin rotation, it would then take two rotations for it to complete a full tumble and return to its original position.

Quantum Theory ascribes this later mystery to the existence of an unusual *spin-space*. Bruce Schumm asks, "So the question is, what exactly is spin and this oddly construed spin-space in which it lives?" He subsequently answers, "We don't really have a clue about the physical origin of spin."[14] This is perhaps another example of Quantum Theory making the mathematics work, while being challenged to interpret the physical phenomena. Again, these mysteries are better understood if quarks and electrons are recognized as composites rather than elementals.

The Origin of Particle Mass

Penergy does not have *mass* as we define it until it is spun into a blackhole or particle, which then gives that Penergy content, definition, and location. Particle mass is much different than blackhole mass, which will be discussed in Chapter Seven in the discussion of *Dynamic Gravity*. Mass is one of the core characteristics of a particle, and according to Einstein is a measurement of the amount of Penergy comprising the particle.

Because of Einstein's famous equation E=MC² and other related equations, mass is often expressed in terms of energy and vice versa. Physicists use a single term called the *electron volt* (eV) as a measurement for both mass and energy. An electron volt is the amount of energy gained by an electron moving across a one-volt, electric field.

To be consistent with current scientific terminology of which many readers may already be acquainted, the eV will be used to express mass. MeV stands for a *million electron volts*, which is the typical value used for most particle masses. If the reader is challenged by the term electron volt, don't worry, it's only a measurement term. Think of it like other measurement terms such as an *inch* or a *pound* that are used simply to state a value. Here the term is used for comparative purposes and its exact value is unimportant.

The Standard Model posits the existence of a special field called the *Higgs Field*. It is named after Physicist Peter Higgs who was instrumental in creating the theory for its existence. It is a field that supposedly exists throughout the universe and gives mass to particles. In other words, it is the particle's interaction with this undetected, hypothetical field that causes particles to have mass.

The field is theoretically created in conjunction with the Higgs boson particle. This boson is a very large (*125,000 MeV*) particle that has not actually been seen but is believed to have been brought into existence at the Large Hadron Collider (LHC). Inside the LHC it survives for only 10^{-22} of a second.[15] That is less than a millionth of a millionth of a billionth of a second. The creation of the Higgs boson is a rare occurrence happening within the LHC only once in every ten billion particle collisions.[16]

Again, the Higgs boson has not actually been seen. Its existence is implied by reading the decay debris detected in the collider.[17] Reading debris created in particle colliders is notably imprecise. The process has been compared to throwing a piano out the window and then trying to determine all the piano's properties by analyzing the sound of the crash.[18]

The Higgs boson was initially predicted to be much larger but turned out to be multi-billion times too light.[19] For the calculations to come within the vicinity of the mass observed, the mass number had to be *fine-tuned (think mathematically manipulated)*.[20] Physicist Ben Still tells us that the lack of anticipated size of the Higgs Boson, "...is a big argument against the Standard Model and is known as the Hierarchy Problem."[21] Despite these problems, the Higgs boson was reportedly discovered in 2012 at the LHC.

The Higgs boson supposedly maintains the Higgs field, yet the boson decays very quickly, so it does not naturally exist in our universe. Particles theoretically gain mass by interacting with the otherwise undetectable and ubiquitous Higgs field in a way that slows the movement of the particle, which causes it to act like our everyday understanding of mass. According to Physicist Lawrence Krauss in his recent book *The Greatest Story Ever Told ... so far,* if the Higgs field can make a particle more sluggish, the particle *acts* like it has a commensurate amount of mass.[22]

The problem with sluggishness equaling mass is that it is inconsistent with Einstein's $E=MC^2$. According to Einstein the mass of a body is a measurement of its energy content, meaning pure energy content. Apparently in the Higgs field the particle does not actually possess the requisite pure energy represented by its mass, it only *acts* like it does. There may be some mathematics to support the theory, but apparently there is no actual evidence the Higgs field exists, except by the presumption that if a particle exists then a corresponding field exists.[23]

While the hypothetical Higgs is widely believed to exist, some physicists have reservations about it. Physicist Harald Fritzsch tells us that physicists invented the hypothetical Higgs field simply out of the need for a theory to give particles mass.[24] Astrophysicist David Lindley tells us the Higgs field explanation for mass is ingenious, but it is a trick, a gadget, a kludge, as computer programmers would say of a piece of code that is tacked on to a piece of software to perform some necessary but overlooked task.[25] Physicists Don Lincoln tells us, "…the proposed Higgs field did not arise from any fundamental principle. In a sense, one can think of the Higgs proposal as a band-aid that patched up a nasty wound in the electroweak theory."[26] Lawrence Krauss tells us physicists don't have the slightest idea why the Higgs boson and field have the properties they do. "It seems completely ad hoc – a convenient addition to the theory that makes it work and allows the physical world to exist…without any underlying compelling mathematical rationale." [27] Obviously there are some serious concerns about the validity of the Higgs scenario.

The Higgs boson might be everything the physicists say it is, but producing a knot of medium-light-density Penergy by slamming protons together will always produce a big, unstable particle because that is how Penergy works. Predicting a big particle will decay into smaller particles in a certain format is not convincing evidence for a particle producing an undetectable field that somehow gives particles mass. Slamming high energy particles together to produce a knot of Penergy and a cascade of lighter particles only proves the presence of our thin-density Penergy, and the instability of heavy particles created within it.

The Higgs theory may satisfy a theoretical need, but it is difficult to believe and not foolproof, so particle mass could just as well be attributed to Penergy density or some other particle dynamic. Knowing exactly how particles got their mass may not be essential to our building a new vision of the early universe, but I've always disliked the Higgs approach to mass and have pondered the issue at some length. Let's speculate on a possible alternative theory.

An Alternative Theory for Mass

Let's start with two associated observations regarding the nexus between mass, spin, and the spacetime/Penergy medium. Particle interactions have shown particles to have momentum, which implies they are real, take up space, and have mass. Mass is known through our understanding of gravity to curve spacetime. Fast spinning bodies such as blackholes also curve spacetime.[28] If *mass* can curve spacetime, and *spin* can curve spacetime, then perhaps what is actually curving spacetime isn't mass per se but something commensurate with mass, such as the *spin* of the particles making up that mass.

Tors are the only particles spinning on their own axes and being pure energy presumably do so at a *very* high rate. Things that spin on an axis at a high rate create a *gyroscopic effect*. That means that the spinning body is inclined to stay in the same position relative to its spin axis, resisting any movement or tilt to that spin axis. You may recall this phenomenon if you have ever played with a toy gyroscope. Physicist Frank Wilczek describes elementary particles as ideal, frictionless gyroscopes that never run down. He also points out that the faster a gyro rotates, the more effectively it will resist attempts to change its orientation.[29]

The measurement of a particle's mass refers not to the amount of matter that the particle is made of, but rather to the particle's inertia, or the degree to which it resists being accelerated by a force.[30] For a spinning body, that resistance to change refers to its axis orientation and is called *orientational inertia*. The quantity that measures orientational inertia is called *angular momentum*.[31] So, the more angular momentum a particle has, the more resistance there is to a change in its orientational inertia.

Perhaps a body's resistance to a change in speed or direction, which is how we measure a body's *inertial mass*, is simply the resistance inherent in many Tors acting like mini gyroscopes. The more spinning Tors that make up the body, even in all sorts of orientations, the more resistance to movement and hence the more measurable mass. Only in a situation where all the spinning Tors are perfectly aligned and balanced would one find a *massless* body such as in the configuration of a photon, which will be discussed in Chapter Six.

More on Mass

This alternative theory for mass makes sense, but there still seems to be more to mass than we currently know. For instance, the mass of a proton is measured to be about 938 MeV, whereas the mass of each of the three quarks making up the proton is estimated to be less than 5 MeV. There are many theories for how this could be, but none of them is compelling. Likewise, the mass of a Pion is measured to be about 140 MeV, but it is comprised of only two quarks each with a mass of less than 5 MeV.

Perhaps it is the state of the quarks that makes it difficult to determine their mass. If quarks are independent particles and have freedom of movement, we should be able to measure their mass fairly accurately. But if quarks are part of a long chain of particles, their freedom of movement would be limited and nailing down their mass would be a challenge. So, perhaps it is the level of physical freedom of the quarks we don't understand. Physicists Ben Still reminds us that the quarks within protons are confined to a tiny space, so there is a large uncertainty to their momentum and energy.[32] There is obviously more to the mass, energy, and freedom of movement of quarks yet to be discovered.

The three quarks in a nucleon normally appear with two of their spins aligned and one quark with its spin in the opposite direction. When that one odd quark flips over, resulting in all three quarks having their spins aligned it is called a delta quark, which otherwise has the appearance of any other nucleon. We are told by Physicist Timothy Paul Smith that the flipping of the third quark causes the proton's mass to jump from 938 MeV to 1232 MeV.[33] It is unknown why this happens, but it may represent good observational evidence for mass being a function of particle spin dynamics.

Scientists are at a loss to explain why particles have the specific mass they do. In the Tor Model, mass content is not a function so much as an underlying principle, but a consequence of evolution driven by the CAGI. As Physicist Antonio Padilla points out, if the electron were three times heavier, it would have destabilized the atom.[34] That of course is the very reason it's not three times heavier. The underlying principle for mass has more to do with what works from an evolutionary standpoint. If it doesn't work, it doesn't evolve.

> Prediction: Once the viability of Penergy and spinning Tors is recognized, a clever young astrophysicist will create the mathematics to demonstrate the relationship between mass and Tor spin. From there, another clever physicist or astrophysicist will develop the relationship between Tor spin, mass, and the Penergy medium density, explaining why there are three generations of particles.

The above speculation on the origin of mass may not be totally correct, but it brings up a point. This speculation suggests that the idea of mass emanates from a natural attribute of the most elementary particle, the Tor. By building the universe on the premise our Sub-A's simply popped into existence directly out of the speck of energy, the Standard Model has gone down a mathematical path that has passed up opportunities to understand the universe in a more elementary way.

The source of mass might well have something to do with the most elementary particle, but if we don't recognize the most elementary particle, that fact will never be explored. The point is, if the mathematics is taking us down a path of questions without reasonable answers, it is time to stop and look back up the path to see where we may have taken a wrong turn. As observed in the introduction, math can be a wonderful tool, but it is plagued with the potential for wrong turns.

We are proceeding with the hypothesis that our Sub-A's are composite particles composed of Tors and Tryks. Let's examine the idea of composite particles and determine whether they have any virtues supporting the argument for their existence.

Virtues of Composite Elementary Particles

Penergy spinning itself into Tors seems to be the most likely way for particles to have come into existence. Penergy may have spun itself into a pair of complex Sub-A's, but simple elementary particles that combine and evolve into Sub-A's is much more reasonable, especially given the observations and deductions made so far. Many physicists tell us our Sub-A's may have a substructure. Physicist Harald Fritzsch says that though there is no experimental evidence for it yet, it is possible that the electron possesses an inner structure,[35] and that quarks and leptons might consist of smaller building blocks.[36]

Physicists John Barrow and Joseph Silk tell us that physicists would not be surprised to discover ultimately that quarks and leptons have internal constituents.[37] Physicist David Lindley relates that, "Quarks and Gluons may not be truly elementary but built from even more arcane ingredients."[38] Physicist Don Lincoln tells us that, "... it is at least plausible to consider the possibility that quarks and leptons are composed of even smaller particles held within them."[39]

Physicist Jon Butterworth tells us both categories of matter particles, quarks and leptons, may contain smaller constituents.[40] Physicist Kenneth Ford tells us that even the lowly neutrino particle may not be a pure particle, but a mixture of two other particles, each with a definite mass.[41] Chemical Physicist Michael Munowitz tells us that quarks may not be the indivisible, end-of-the-line building blocks, but they are probably only a step or two away.[42]

Stephen Hawking told us that at very high energies we might expect to find several new layers of structure more basic than the quarks and electrons that we now regard as elementary particles.[43] Dr. Jorge Cham and Physicist Daniel Whiteson tell us there is no proof that the particles we see today, electrons and quarks, are the most basic building blocks in the universe, but are [merely] the smallest bits of matter we have seen so far.[44]

According to Astronomer Chris Impey, the Standard Model leaves open a possibility that there's a deeper level of structure.[45] Obviously, though there is no evidence for it yet, many feel the existence of constituent particles comprising our Sub-A's is at least plausible. The following points support the notion our Sub-A's are composite particles.

- As we have seen, composite particles easily account for the missing anti-matter. The anti-matter is not only a part of composite particles, but an essential part. Anti-matter's natural attraction to matter makes it possible for Tryks to exist. As we progress in constructing future building blocks to our universe, the integrated anti-matter particles will again prove to be an essential element in that construction.

- Composite particles could be the underlying source of *magnetism* and *charge*. As discussed in Chapter Four, Tor-Chains with all their spin axes aligned create magnetic fields that encourages other particles to align their spin axes accordingly. As we will discuss in the next chapter, the combination of elements in composite particles will explain charge and why some Sub-A's have differing amounts of charge.

- Composites imply the existence of a constituent core particle spinning on its own axis such as the Tor. The validity for the existence of a particle like the Tor is implied in many of the theories and speculations discussed so far. We have speculated on the spinning of the Tor being the source of mass. We have also speculated further on how the Tor could also be the source of the magnetic, electric, and gravitational fields. It seems like quite a bit to expect from a single particle, but it makes more sense for these attributes to arise out of the most elementary particle than for those fields and forces to have inexplicably popped into existence without foundation or explanation.

- The creation of composite particles infuses natural selection into the evolutionary process, which seems to be a much more credible influence on particle composition than the 'it just happened' approach taken by the Standard Model.

- The Standard Model recognizes particles to be very consistent in their attributes, such as charge and mass, but offers no credible explanation as to why this is so. On the other hand, the idea of evolving, composite particles created through natural selection supports sound reasoning for consistent particle attributes, as discussed above.

- Scientists tell us that all charges are rational multiples of a basic unit of charge. The fact that there are three Sub-A's

A NEW VISION OF THE EARLY UNIVERSE

(*electron, up quark, and down quark*) with the same type of electric charge but with different values makes it probable they share a common, underlying structure. Jorge Cham and Daniel Whiteson tell us that because the charge of the electron perfectly matches the charge of the proton but is opposite, it is another sign there are deeper components underlying those particles.[46]

- As mentioned in the previous section, the mysteries of particle spin and angular momentum are more easily understood if electrons and quarks are recognized as composites rather than as elementals.

All these facets weigh heavily toward today's Sub-A's being composites of more elementary particles, named in the Tor Model as Tors and Tryks. Based on these arguments and what we have discussed earlier we can make our next deduction.

Deduction #12: Given that Penergy spun itself into a cascade of swirls ending in early elementary particles; and given those particles can combine due to magnetic and electric forces; and given the many supporting points for our Sub-A's being composite particles; we can conclude that today's Sub-A's are composite particles made up of elementary particles like Tors and Tryks.

With the conclusion that Tors and Tryks comprise our Sub-A's, we now turn to how that next stage of universal evolution took place.

*   *   *

In this chapter we looked at the conditions of the early cosmic environment and the attributes of early elementary particles. We examined the importance of *natural selection* and the part it plays in assuring each phase and Level of evolution is capable of both surviving within its environment and possessing the capacity to combine again. We gave a definition to *cosmic evolution* as being matter growing in complexity through the Combination and Growth Imperative (*the CAGI*).

We looked at the Cosmic Homogeneity issue and concluded that the current theory called *Inflation* stretches credulity too far and that cosmic homogeneity is more likely due to the cascade effect creating blackholes and particles together, and by the time the SMBHs gained a sufficient gravitational effect, it pulled in the mixture of particles, gases, and particle combinations universe wide, making the universe homogeneous. We also looked at *entropy* and the importance of cosmic recycling.

We next discussed the origin of mass and concluded that the Higgs Field explanation was difficult to believe, and we offered a speculation on an alternative theory. We then examined the virtues of composite particles and concluded that our subatomic particles are most likely composites of early elementary particles. We now turn to see how those composites might be configured.

Chapter Six
The Origin of the Atom

Our new vision of the early universe is well under way and is beginning to look more like the universe we are familiar with today. The supermassive blackholes are taking hold and the intermediate swirls have all but disappeared leaving the skeleton of the cosmic web. The next significant relationship to evolve is between the *fields* and *forces* that serve to bring particles together. The EEPs are combining into stable Tryks, and our Sub-A's are about to form from those Tryks. Some of the intermediate swirls may have consumed enough particle matter to give them sufficient mass to turn them into proto stars.

Tryks brought important emergent qualities to the universe. The fields surrounding the Tryks reflect the fact they are comprised of both Tor-Chains creating a magnetic field and a combination of matter and antimatter creating an electric field. This combination of essential emergent qualities together creates the electromagnetic field, which gave those particles the very important ability to absorb and emit photons.

Photon absorption gave Tryks the capacity for different energy levels to promote various interactions and combinations. These emergent qualities not only assisted the Tryks in combining, but they will prove essential to the growing and combining of all future levels of matter as well.

The cosmic environment remains a hot frenzy of activity, perhaps in a plasma state, too hot for Sub-A's to yet form. But as the universe expanded and cooled, the Tryks would have begun combining. Those new particle combinations would have to be resilient to survive. Many trial combinations of Sub-A's would have been created, testing the combining strength of the Tryk. This slow, testing, transitional phase occurs between all Levels; it is simply a part of the CAGI's evolutionary process. We will proceed as if the Tryks appearing in Figure 4.4 were the dominant form that successfully evolved, but of course the actual configuration is likely much different.

Combining Tor-Chains with Tor singles may have given rise to a problem of how to make those connections stable. This problem may have required a special *connector* to have evolved. We begin our examination of particle configurations with the prospects for such connectors.

Connector Configurations

To smash Sub-A's together at high speed, scientists use large particle colliders such as the twenty-seven-kilometer, circular, Large Hadron Collider (LHC) located underground near Geneva, Switzerland. The collision creates a knot of Penergy big enough to bring into momentary existence large, unstable particles that immediately decay into smaller particles. Scientists have developed detection equipment to record the behavior of these smaller particles, which gives the scientists a basis for theorizing the particle's existence, characteristics, and relationship to other particles.

Some very large particle creations are so fleeting they have not yet been seen but are theorized based on the behavior and characteristics of collision debris. Scientists have identified only four, naturally existing *stable* particles and their anti-particles: the quark, neutrino, electron, and photon. They have also either seen or theorized a host of many unstable particles and anti-particles.

All stable matter as far as we know is made from quarks, neutrinos, and electrons. Photons are considered energy particles that provide matter particles with the requisite energy to move, interact, and combine. Due to shared features, electrons and neutrinos have been placed in a subcategory called leptons. There is also a category for virtual particles called Bosons.

As previously observed, it was the strong Tor field that encouraged a nose to tail coupling creating Tor-Chains and Tryk-Chains. The addition of an opposite-spinning Tor encircling the chain to create a Tryk, however, would likely cause the Tryk-Chain to wobble a bit. Such a wobble could weaken the field connection strength between the Tor-Chains. As the Tryk-Chain links grew longer and the wobbles grew wider, the magnetic field strength of the Tors holding the Tryks together would have grown weaker. The wobble may not have been a problem for short Tryk-Chains but may have been a serious problem for lengthy Tryk-Chains. That wobble might be the source of the small precession value found in all electrons, or the wobble found in the Muon's magnetic moment.

This limitation to Tryk-Chain length could have been overcome if there had evolved another independent particle that could serve as a flexible connector (○) between the Tryks. In this way the Tryk-Chain connection would not be completely dependent on the magnetic force between Tors, allowing for stable Tryk-Chains of any length.

Such connectors would likely be very small and neutral in charge. They might exist without our even knowing it. Being small, neutral, and likely near massless, in the collision debris calculations in colliders and accelerators they could easily be mistaken for neutrinos, or they may even *be* neutrinos. Let's examine the prospect of the neutrino being a connector.

Neutrinos as Connectors

Neutrinos are one of the most abundant particles in the universe. They have neutral electric charge, are nearly massless, and seldom interact with other matter.

In the Tor Model, Neutrinos may serve as a connector particle for electrons and positrons. Neutrinos only seem to appear in conjunction with an emission of an electron or positron due to particle decay or during the fusion process at the heart of a star. Every time an electron or positron is spit out, so is a neutrino. This close tie suggests a possible physical connection between the particles.

The neutrino may be a *Majorana* – a particle that is its own anti-particle. If the neutrino is its own anti-particle, it would have a unique configuration. As such, it may serve as a universal connector allowing for the connection between all sorts of Tryk-Chains and other particles.

If mass is the resistance to a change in velocity due to the gyroscopic spin of its Tors as discussed in Chapter Five, it would mean that mass is a function of particle configuration. Mass would be commensurate with Tor content except where the configuration allowed for a perfect balance in the gyroscopic effect of its Tors, as suggested in the description of the photon (*illustrated in Figure 6.6*). If that's the case, it may explain how the neutrino can oscillate - change mass/identity while in flight. The oscillation would mean the neutrino simply changed in configuration. That would mean, of course, that the Tau, Muon, and Electron Neutrinos are all the same particle with changeable configurations. For instance, the neutrino pictured in Fig. 6.1 could, while in flight, pick up an electron at one end and a positron at the other, changing its configuration and mass content without changing its neutral charge or other characteristics.

Let's speculate on how a neutrino might be configured. As illustrated in Figure 4.3c, the four Tor-chains with different lengths comprising a bundle would be drawn to and rotate around each other due to electric field attraction. If the chains were joined by a single Tor encircling the end of the bundle, it would create an electrical imbalance that would allow it to connect to either a positive or negative Tryk-Chain. The connection could be due to either the respective fields joining, or the single Tor being shared between the connector and the Tryk-Chain, like electrons are shared by atoms to create molecules.

The ends of the connector would be capable of attracting either a left or right spinning Tor, making it capable of joining with either a positive or negative Tryk-Chain. The bundle's attraction of a single Tor would create an electric charge, which would attract photons that when absorbed would change the neutrino's energy and momentum, giving it the attributes of a more massive particle without changing its charge value or other characteristics. Having both left and right spinning Tor-chains exposed at its ends means the neutrino could attach to either kind of spinning Tor-chain, making it an ideal, flexible, universal connector.

Neutrinos are known to only spin to the left, strongly suggesting they are a Majorana – their own anti-particle. The neutrino in Figure 6.1 is moving upward as shown.

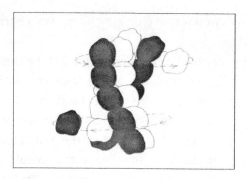

Figure 6.1 This flexible configuration would allow the Neutrino to serve as a connector between left or right spinning Tor/Tryk-Chains.

I don't mean to suggest this configuration is the exact configuration of the neutrino. I am once again only demonstrating how Tors and Tryks as building blocks could result in all kinds of particle configurations and be used to answer some of the mysteries currently existing in particle physics.

Quark Configurations

According to the Standard Model, protons and neutrons are composite particles comprised of three quarks and a sea of virtual *gluons* that by exchange hold the quarks together. Quarks and gluons have not been isolated or observed directly but are theorized to exist based on experimental evidence.[1] Quarks are believed to be elemental, are never found as a single unit, and are normally found paired together in twos or threes. They are believed to carry electric charges but only in 1/3rd or 2/3rds of the whole 3/3rds charge value of the electron.

Quark Charge Values

In the Tor Model, quarks are not elemental but are comprised of Tryks, and their fractional charges can be easily constructed from those Tryks. The strength of the electric force is measured in *charge value,* but no particle possesses the inherent attribute of *charge*. Charge is simply a measure of the strength in the *interaction between the fields* of Tors. The greater the differential in the number of Tors, the greater the strength of the charge. To measure the differential, we can place a theoretical value on each Tor, but again the particle does not actually possess an intrinsic charge, only a relative measurement value based on the overall attraction/repulsion between the Tor fields.

As observed, the measure of quark charge is $1/3^{rd}$ and $2/3^{rds}$ of the electron. For the sake of demonstrating how these partial electric charges are possible at the Tryk level, let's give each individual Tor particle a hypothetical $1/9^{th}$ charge value. Accordingly, right-spinning (*negatively charged*) Tors would have a $-1/9^{th}$ charge value, and a left-spinning (*positively charged*) Tor would have a $+1/9^{th}$ charge value. Again, these values are ascribed for demonstration purposes only.

With these values and terms, we can construct all sorts of particles with all sorts of electric charge values both positive and negative. Because Tryks contain both positive and negative Tors, each pair of which would add to zero charge, we only need to calculate the charge value based on the *net number* of charges.

To get the net number of charges for each Tryk we add up all the charges. As shown in Figure 6.2a, a Positive Tryk would have a gross charge value of $+4/9^{ths}$ for the four positive (*left spinning*) Tors, and $-1/9^{th}$ for the one negative (*right spinning*) Tor, giving it a net charge of $+3/9^{ths}$, or a $+1/3^{rd}$ overall charge value.

As shown in Figure 6.2b, a Negative Tryk would have a gross charge value of $-4/9^{ths}$ and $+1/9^{th}$, giving it a net charge of $-3/9^{ths}$, or a $-1/3^{rd}$ overall charge value. In summary, the *electric charge* value of any particle is determined by the net differential in the positive and negative Tor count making up the particle.

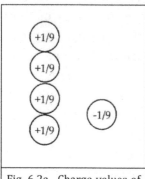 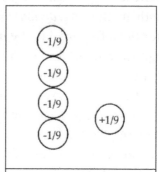

Fig. 6.2a. Charge values of a Positive Tryk: net +3/9ths or a +1/3rd overall charge.

Fig. 6.2b. Charge values of a Negative Tryk: net −3/9ths or a −1/3rd overall charge.

Figure 6.2 Charge values of Tors added together to give a *net charge* value of a Tryk.

The values given above are arbitrary and are used only for the sake of example. It is not my goal to prescribe a specific charge value for the Tor but only to show how charge values *might* be distributed between the applicable particles.

Deduction #13: Given the arguments against the source of electric force being virtual boson exchange; and given the simplicity of that source being the natural attraction/repulsion between the fields of Tor particles; we can conclude that the source of the electric force is a differential in the number of positive and negative Tors making up a particle: an abundance of positive Tors gives the particle a positive charge, and vice versa.

Prediction: An enterprising young physicist or astrophysicist will someday win a special prize for extrapolating the mathematics of quantum field theory into a theory of the magnetic and electric forces conveyed through the fields of early elementary particles.

Simple Quark Configuration

Using these Tryk charge values we can now easily create a composite particle with an up-quark charge value of +2/3^{rds} by joining two positive Tryks into a single chain as illustrated in Figure 6.3a. The anti-particle of this simple illustration of an up quark would have the mirror image configuration and carry a −2/3^{rds} charge, as illustrated in Figure 6.3b.

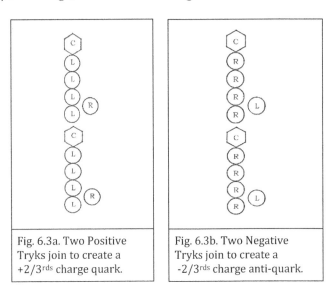

| Fig. 6.3a. Two Positive Tryks join to create a +2/3^{rds} charge quark. | Fig. 6.3b. Two Negative Tryks join to create a −2/3^{rds} charge anti-quark. |

Figure 6.3 Representations of a simple quark configuration and its anti-particle.

Complex Quark Configuration

A quark configuration might be as simple as Tryks hinged together by a connector, but it is possible that out of the hot frenzy evolved much more complex particles. A down quark for example might be made up of a chain of electrons, positrons, neutrinos, universal connectors, and positive and negative Tryks. An example is illustrated in Figure 6.4.

This configuration would have a net −1/3rd charge value, which is the exact charge value of the down quark. Should it

THE ORIGIN OF THE ATOM

spit out the **Electron/Neutrino** couplet at its corner, it would then have a +2/3^{rds} charge value, becoming an up quark, which is the process changing a neutron to the proton. Likewise, should this up quark spit out the **Positron/Neutrino** couplet at its corner, it would then have a -1/3rd charge value, becoming a down quark, changing a proton to a neutron. The ease at which particles decay into other particles in this fashion makes the argument for composite particles compelling. This down quark configuration is not definitive and is only one of many possible configurations for a quark made from Tryks.

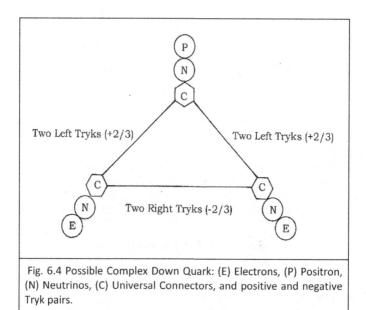

Fig. 6.4 Possible Complex Down Quark: (E) Electrons, (P) Positron, (N) Neutrinos, (C) Universal Connectors, and positive and negative Tryk pairs.

Figure 6.4 This Quark configuration allows for a simple decay to convert to either an Up or Down Quark.

The Simplest Quark Configurations

In keeping with Occam's razor, let's see how a minimum quark configuration might work. A decayable +2/3^{rds} up quark would only need at a minimum two positive Tryks, an electron, positron, and accompanying neutrinos: a +2/3, -1, +1 combination. If a decay resulted in the removal of the +1 positron, the result would leave a -1/3rd down quark. Likewise, if we started with a decayable -1/3rd down quark, we would only need at a minimum one negative Tryk, an electron, positron, and neutrinos: a -1/3, -1, +1 combination. If a decay resulted in the removal of the -1 electron, the result would leave a +2/3^{rds} up quark.

If the quark decay was initiated by the absorption of a high energy photon that immediately decayed into an electron-positron pair, we would have all the ingredients necessary for the quark decay. The decay would leave a Tryk-Chain connected to either an electron or positron, while ejecting an electron or positron and its neutrino connector, depending on the type of decay.

Again, I am not suggesting that any one specific configuration is how quarks are comprised. Obviously, there are many options available, and the configurations discussed are only to demonstrate some possibilities. With some head scratching, new experiments, and a dab of mathematics, an ingenious physicist or cosmologist will someday figure it all out.

Although many different Tryk combinations would have been created, natural selection chose our up quark with a +2/3^{rds} charge and down quark with a -1/3rd charge to dominate. These quarks would have become dominant due to their need to be in threes to be stable, and to combine their partial charges to add up to the +1 charge (*+2/3, +2/3, -1/3*) of a proton, equaling perfectly and offsetting the -1 charge of the electron. Those configurations eventually allowed the hydrogen atom to form.

Quark Evolution

That sequence of evolution was a lot to take in so allow me to rephrase it for clarity. In the Penergy medium frenzy, many complex quarks with various charge values would have been created. But not until a stable, three-quark combination, with a +1 net charge evolved did natural selection allow the up and down quark configurations to become dominant. But how did the universe know it needed a +1, positively charged particle?

It was not a conscious need but a need for cosmic evolution to progress. The CAGI described in Chapter Five compels matter to combine until it creates a new Level of matter with its own capacity to combine. Time is of no concern. The CAGI will take as long as necessary and create as many trial combinations as necessary for a combination to come about that qualifies as, or is a sub-stage to, the next stable Level of matter.

In this case, quark-like combinations and electron-like combinations would be created over and over until configurations capable of combining into a larger, stable particle came about. Since the electric force was now available to bring particles together, it makes sense that the primary attribute driving the evolution of these particles was the attraction of their opposite spinning Tors and Tor-Chains. Once the stable electron with a -1 negative charge became dominant, it was only a matter of time before a Tryk combination came about to precisely off-set that charge value and create the proton.

Meson Configurations

According to the Standard Model, the nuclear force binding the protons and neutrons together to form atomic nuclei is due to the exchange of a meson called a Pion.[2] A meson is comprised of a quark and anti-quark. In the Standard Model, the mirror-image quarks contain a different *color charge*, so they don't annihilate. A meson can also be comprised of a quark and anti-quark from different generations. Mesons are, however, unstable and suffer an immediate decay when created in a shower of collider particle debris.

The Tor Model recognizes the momentary existence of mesons created from various quarks from various generations. The presence of a color charge may not always be necessary, however, as there are many combinations of a quark and anti-quark having different spin signatures that can combine without annihilation. Consequently, the Tor Model offers no specific configuration for the two-quark meson but recognizes the importance of its evolutionary development in being instrumental in holding the nucleus together.

Electron Configuration

According to the Standard Model, the electron is an elementary particle without constituent parts. Frank Wilczek reports in his recent book, *Fundamentals – Ten Keys to Reality*, that electrons can be fractured into smaller particles with partial charges through the *fractional quantum Hall effect*. But these are not real particles but only elements that act like independent particles, called quasiparticles. They are created only under precise laboratory conditions.[3] This research, however, suggests the possibility that electrons could have an inner structure, even if it's only manifested under a very limited and extreme experimental environment.

In the Tor Model, the electron is a composite particle made up of Tryks. The key element in the configuration of an electron is that it carries a whole $9/9^{ths}$ negative-charge value, which means that using the values we established already it has a differential of nine negative Tors. Three negative Tryks in a chain would fulfill the requirement as illustrated in Figure 6.5a. The positron (*anti-electron*) would of course have the mirror-image configuration as illustrated in Figure 6.5b. Other Tryk and particle combinations comprising an electron are of course possible.

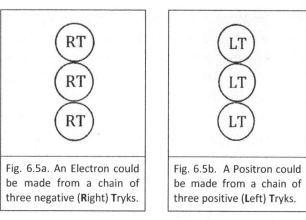

Fig. 6.5a. An Electron could be made from a chain of three negative (**R**ight) Tryks.

Fig. 6.5b. A Positron could be made from a chain of three positive (**L**eft) Tryks.

Fig. 6.5 Possible configurations of an Electron and Positron.

The electron illustrated above is made from a *chain* of Tryks. One might consider creating an electron from three unconnected, independent Tryks. That would seem equally feasible, but it is not possible due to the *Pauli Exclusion Principle*. That principle says that no two like fermions (*electrons or quarks*) can together occupy the same state, meaning carry on the same function. We need not go into the principle further but know that if comprised of three Tryks, they are more likely in a chain rather than independent units, unless each unit possesses a unique trait giving each a slightly different identity.

In the chain configuration, the three, single, rotating Tors can co-exist because they are not rotating in the same plane and therefore are not in the same state. The Tryks being in a chain is probably the reason the electron is thought to be elementary since the chain of Tors and encircling Tors comprising the Tryks would have a very strong coupling capacity and high binding energy, making them difficult to break apart.

Even without the existence of a connector, three Tryks may have been able to join nose to tail to create a strong chain. We have concluded that due to the very strong field strength of the Tor, in the early universe the magnetic force could reasonably overcome the electric force if the Tors came together nose to tail. We have also observed this to be the case in experiments involving two otherwise repulsive electrons coming together nose to tail to create a combination called a Cooper Pair.[4] This phenomenon came to light when scientists were exploring theories of superconductivity (*the flow of electricity without resistance*). Again, it only happens in extreme laboratory conditions. We know, however, that the strength of forces bringing particles together can change under different energy conditions, and the early universe certainly provided the opportunity for different energy conditions.

Boson Configuration

According to the Standard Model bosons are whole-integer-spin particles and are mostly virtual particles that are theorized to facilitate the absorption, emission, creation, and decay of Sub-A's. The group is comprised of photons, gluons, gravitons, W and Z Particles. Being virtual means they exist only momentarily, flitting in and out of existence. Their hypothetical existence requires them to borrow energy from the vacuum of space, but as shown earlier the vacuum apparently does not possess the requisite energy to support their number.

The Tor Model recognizes their possible existence but believes their importance in our universe is uncertain. The Tor Model has an alternative explanation for the source of force and how particles interact and decay, so the possible makeup of these particles will not be addressed. Photons are known to also be stable particles, so let's examine how they may have evolved.

Photon Configuration

Photons are widely known as *light*. They are the particles that are absorbed and emitted by matter at specific wavelengths, which then enter our eyes creating impulses that are interpreted by our brains giving us vision. Photons do not naturally interact with each other. Except when passing through a clear medium, such as air, glass, or water, they travel in a straight line and at light speed, roughly 3.0×10^8 meters/second.

Photons are never at rest, so there is no way to measure their rest mass, but experiments indicate they are massless. They can be *absorbed* and *emitted* by charged particles, e.g., quarks and electrons, but otherwise exist only in constant motion within the Penergy medium. When absorbed, their energy is transferred to the particle, boosting its *kinetic energy* (*energy associated with motion*).

Photons can have very high energy and can decay into most any kind of matter-antimatter pair. They can do this while remaining massless, making their composition extraordinary. They may or may not be comprised of Tors. If not, it would mean the early Penergy also spun itself into an entirely different kind of stable particle that is peculiar to the photon. We will proceed with the notion photons are comprised of Tors, but other options may be equally credible.

The photon is so extraordinary that trying to guesstimate its configuration is near folly, so don't rely on the description that follows as golden. I have only modest faith in it myself, but since I have done the work to visualize something that works, I may as well give it to you with the understanding that other options are certainly possible.

To be charge neutral, photons must have an equal number of left and right-spinning Tors. Having little or no mass means their Tor spin is in perfect, or near perfect, balance in order to have little if any gyroscopic effect. Having no mass, photons do not distort the surrounding Penergy medium, thus having no gravitational effect on other particles.

Photons do have a measurable momentum and are known to have the effect of an impact when absorbed or scattered off another particle. That means the photon possesses a quality we call mass though experimentally it shows itself to be massless. This is easily understood if the photon is configured with Tors without the Tors demonstrating any measure of mass by having their spin axes in perfect balance.

All these characteristics might be accounted for if the photon is comprised of a loop of both left and right Tor-Chains with equal length differences, as illustrated in Figure 6.6. The loop length and number of chains pictured are merely for example. This multi-strand loop configuration is easy to construct being comprised of simple Tor-Chains. It can be easily assembled from a knot of Penergy produced by annihilation, and easily broken apart and reconfigured to produce any matter-antimatter pair. This design also satisfies the fact that the mirror image of a photon looks the same as the original, making it its own anti-particle.

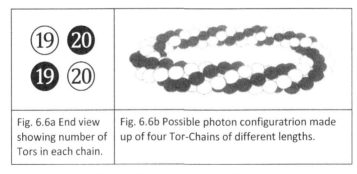

| Fig. 6.6a End view showing number of Tors in each chain. | Fig. 6.6b Possible photon configuratrion made up of four Tor-Chains of different lengths. |

Figure 6.6 Possible configuration of the photon comprised of equally paired opposite-spinning Tor-Chains.

The photon is believed to carry both the electric and magnetic fields and to be the source of the combined *electromagnetic field*. Each of the Tor-Chains spin causing the loop to twist into an elongated knot giving it a long axis. Photons spin on their long axis as the loop twists through space. In a beam of photons with their spins aligned, the angular momentum of the individual photons adds up and the beam imparts a measurable torque or twisting force.[5] This suggests that photons do indeed spin on a long axis. The loop traveling on its long axis assumes the shape of a long Tor-Chain that naturally creates a *magnetic field*, as illustrated in Figure 6.7.

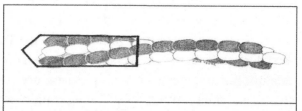

Fig, 6.7 A twirling photon loop assumes the shape of a long Tor-Chain, naturally creating a magnetic field.

Because the photon spins on a long axis, the rotating Left and Right Tor-Chains reflect a differential in Tor counts, mimicking an *electric field* as illustrated in Figure 6.8.

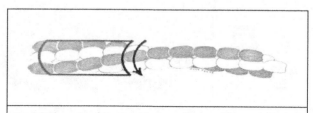

Fig. 6.8 The spin of the Tor-Chains reflects a differential in Left and Right Tor count, inducing an electric field.

The electric field in turn induces and reinforces a magnetic field that extends perpendicular to both the electric field and direction of travel. Together they produce an *electromagnetic field,* as illustrated in Figure 6.9. Because photons are so ubiquitous throughout the universe, the induced EM field is likewise ubiquitous.

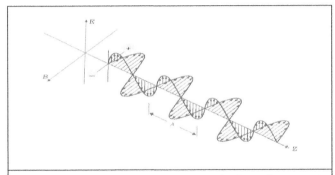

Fig. 6.9 The electric field (E) extends perpendicular to the magnetic field (B), and together they create the electromagnetic field.

The description of the photon's movement also serves to describe the relationship between magnetism and electricity. Scientists have long known that electricity moving through a wire induces a magnetic field. This makes perfect sense if one imagines the spinning electrons moving through the wire headfirst, parallel to their axes of spin. All of those moving, like-spinning electrons in a head to tail arrangement would mimic a Tor-Chain and therefore induce a magnetic field.

Likewise, consider a generator. When a magnet comprised of many particles with aligned spin-axes is rotated, such as an armature, it creates an unequal balance of charged particles, mimicking and inducing an electric field, causing electricity to flow in the adjacent wires. Each of the fields in turn induces the other.

The photon is said to be absorbed/emitted by a charged particle, e.g., a quark or electron. We do not know how this is done. Perhaps the attachment/absorption is due to the charge differential in the charged particle attracting and holding onto the mimicked electromagnetic field of the photon. In any event, the attachment changes the particle's kinetic energy. The photon seems to disappear at the time of absorption. Perhaps it has become a part of the particle or is simply spinning undetected around the particle.

Once stable quarks came into being, the CAGI could begin making combinations in its effort to offset the stable, negatively charged electron, eventually creating protons and neutrons.

Evolution of a Nucleus

The Standard Model and Tor Model agree on the basic configuration of the proton being comprised of two up quarks and one down quark. Those quarks have charges of $+2/3^{rds}$, $+2/3^{rds}$, and $-1/3^{rd}$ respectively, giving the proton a +1 net charge that perfectly offsets the -1 charge of the electron. This allows for the eventual creation of a neutrally charged hydrogen atom composed of one proton encircled by one electron. The reader may recall from high school science class that each element on the periodic table is distinguished by the number of protons in the nucleus, and that two or more protons, because they carry a positive charge, will naturally repulse each other. Quarks being held together by sharing gluons theoretically spill over to protons sharing gluons, overcoming their natural repulsive force. Apparently overcoming that repulsive force between protons was insufficient for the gluon exchanges alone, necessitating the involvement of another particle - the neutron. Let's look at how a neutron and the nucleus may have evolved.

According to the Standard Model

According to the Standard Model, the following points theorize protons and neutrons. They were created within the first minutes of the big bang and came together to create nuclei during a period/process called Primordial Nucleosynthesis. Unconfined neutrons decay in less than fifteen minutes. In the initial high energy conditions following the big bang, protons and neutrons could exchange identities converting back and forth into each other, but these conditions were short lived, coming to an end as the universe expanded and cooled. Once the conversion stopped, the universe had only fifteen minutes to create the nuclei, otherwise it would have run out of neutrons making it unable to create atoms larger than hydrogen.[6] The protons and neutrons immediately began combining to create nuclei with different configurations, primarily hydrogen, helium, and some of their isotopes. All the existing neutrons were taken up into nuclei within a few minutes of the big bang.[7]

This sequence is part of the Hot Big Bang Theory. That theory may be feasible mathematically, but it makes more sense for complex particles to have evolved from something smaller, and for each aspect of evolutionary development to follow from a reasonable cause without the need of anything simply popping into existence. Let's continue with the story of how nucleons evolved, but from a more natural, evolutionary sequence.

According to the Tor Model

A neutron is comprised of two down quarks and one up quark, giving it a neutral (*-1/3, -1/3, +2/3*) net charge. Up quarks and down quarks are similarly configured with the down quark having a little more mass. As we saw in our discussion on quark configuration, it is relatively easy to create a neutron by adding an electron/neutrino couplet to an up quark.

Protons and neutrons can exchange a part of themselves, a two-quark meson called a pion. Pion sharing allows the protons and neutrons to rapidly exchange identities, which also helps to overcome the natural repulsive forces between the protons.[8] Neutrons apparently would have to evolve to become a neutralizing agent in the nucleus of atoms larger than the hydrogen atom. Presumably this was necessary to give the nucleus more product from which to make pion exchanges.

Given that neutrons naturally decay into protons in a matter of minutes, it makes more sense for neutrons to have evolved *after* the hydrogen atom evolved, which allowed neutrons to readily join a nucleus before decaying. Once inside a nucleus, neutrons become quite stable. It makes no sense for the neutron to have evolved before the hydrogen atom because without the need for the neutron in atoms larger than hydrogen, the neutron would have no reason to evolve. There would have been no need to drive its evolution.

Evolution of the Atom

Once the four stable subatomic particles (*photon, electron, neutrino, and quark*) evolved, the next Level of evolutionary development - the hydrogen atom - probably went smoothly but not necessarily quickly. After the photon evolved from simple Tor-Chains, the next stable particles were probably made up of Tryk-Chains. Evidently the Tryk combination creating the electron was quite stable and naturally selected early. To be instrumental in the next Level of matter, its negative -1 charge needed an offsetting positive +1 charge particle.

The anti-electron (*positron*) has a positive +1 charge value, but because it is the mirror image of the electron they annihilate, making it unsuitable for the off-setting task. Ultimately it was from the evolution of various quark combinations with $1/3^{rd}$ and $2/3^{rds}$ charge values that the proton with a positive +1 charge value evolved. The proton could then combine with the electron creating the hydrogen atom.

From there, neutrons would evolve having nearly an identical configuration as the proton. Neutrons gave product and stability to a nucleus allowing heavier atoms to evolve. The hot environment and high energy pion exchange forces could bring protons and neutrons together to create the larger nuclei of helium, lithium, and beryllium. No other particle combinations were necessary to create the atomic level of matter.

The atom brought an exciting emergent quality. Looking at any single atom, one would not expect it to have any special attributes, but when combined with other atoms, interesting things happen. One would not expect the simple combination of two hydrogen atoms and one oxygen atom to produce much, but in bulk it produces one of the most important elements in our universe – water. As you know, various combinations of atoms produce all the other elements in the form of gases, liquids, minerals, and metals that comprise us and our environment. Atomic combinations are truly wonderous emergent qualities, yet they pale in comparison to the emergent qualities that evolve in future Levels of matter.

The Sequence of Particle Evolution

All particles are created from Tors.
Individual Tors → Tor-Chains → Photons
Individual Tors + Tor-Chains → Tryks → Tryk-Chains
Tryk-Chains → Electrons & Quarks & Neutrinos
Variously Charged Quarks → Protons
Protons + Electrons → Hydrogen Atoms
Proton + Electron + Neutrino → Neutron
Protons + Neutrons + Electrons → Helium & larger Atoms

Because the universe was relatively large before particles were created, the mass-energy and temperature scale of the early universe was high, but perhaps not nearly as high as prescribed by the Standard Model. The combining and building in complexity of all these particles likely transpired using naturally occurring conditions and forces, without the need for ultra-high temperature, though at times the early universe would have been very hot.

In the Standard Model's vision of the universe, the next step in its evolution commensurate with the creation of atoms was the liberation of photons creating the Cosmic Microwave Background (CMB).

The CMB

According to the Standard Model, when the early universe eventually cooled to 3000K that allowed electrons and protons to combine to form atoms, it freed photons that had previously been bound up in those particles. This sudden liberation of photons sent them soaring throughout the expanding medium. Most of those photons are still flying along untouched.

The expansion of the universe stretched the wavelengths of that thermal radiation causing it after 13.8 billion years to lose energy and cool down to just a few degrees above zero. The remnants of that event can be seen today by the presence of a background of very cool 2.75K microwave photons that uniformly exists in every corner of the universe. Those remnants are called the *Cosmic Microwave Background* or the CMB. Their presence is one of the cornerstones of the big bang theory and one of the yardsticks for measuring both the age of the universe and the rate of its expansion.

Figure 6.10 The Cosmic Microwave Background; provided by coolcosmos.ipac.caltech.edu. The light and dark areas show a very slight temperature difference hypothetically due to fluctuations in the conditions of the primal universe.

The Standard Model's hypothesis for how the CMB photons lost their energy is reasonable and may be correct, but the Tor Model offers a possible alternative explanation. The stretching of the wavelengths of the CMB photons may not have been caused directly by the expansion of the universe, but by the gradual loss of cosmic Penergy density during their journey. The original CMB photons were likely the first-generation photons that evolved from Tor-Chains before more complex matter particles evolved. Chapter One discussed the phase transitions that caused mass reductions in the first two generations of matter particles. The continuous reduction of cosmic Penergy density would have caused the first generations of photons to shed some of their mass content, as well, giving them less energy and a longer wavelength.

The mass that was shed was not simply a percentage or random amount but the exact amount that allowed the residual photon to continue to exist in the thinning Penergy Medium. This reduced all photons to the same energy level. The amount of Penergy density would have been consistent throughout the universe, so we would expect to find the energy levels of those early photons to be very consistent, as well.

The universe was much smaller when the SMBHs and CMB photons were created. Though all the SMBHs may not have been fully developed, by the end of the cascade era they may have had sufficient overall gravitational influence to require the photons to lose energy climbing out of the surrounding gravitational fields for much of the early part of their journey, which would have caused some slight variation in the photon energy content.

The Standard Model places the time for the creation of the CMB at 380,000 years after the big bang. Although the Tor Model's timing will work out to be different, the creation of the CMB appears reasonable in both models.

In their book *Cosmology for the Curious,* Delia Perlov and Alex Vilenkin tell us that one of the problems with the Standard Model is that to account for the existence of clusters and super clusters of galaxies, scientists must postulate the existence of small density fluctuations which gradually evolve into those structures. But they have no answer for the origin of those small density fluctuations.[9]

In the Tor Model those fluctuations could be caused by the gradual disappearance of the intermediate Penergy swirls that left large holes in the galactic structure. Those swirls would have been disappearing at the same time particles were being created and perhaps even up until the time atoms were created. Those gradual changes in the galactic structure may have created the conditions that we interpret today as fluctuations in the primordial universe. Of course, in the Tor Model such fluctuations are seemingly unnecessary because the creating of clusters and superclusters of galaxies is accounted for in the Cascade Effect described in Chapter One.

* * *

The premise of the Tor Model is that particles, fields, and forces did not inexplicably pop into existence but naturally *evolved* from the conditions of the early universe. Having established their evolutionary origins, we then turned to how early elementary particles could have combined to create our Sub-A's. Since we cannot know the exact makeup of our Sub-A's from those particles, we put together plausible configurations. Again, these configurations of quarks, electrons, neutrinos, and photons are put forth only as an example of how those composite particles *might* be configured. The only certain attribute is that they are all comprised of a single early elementary particle like the Tor.

In our exemplar configurations, all particles are comprised of Tor and Tryk-Chains that can easily be broken down and quickly reconfigured, which accounts for how quarks, leptons, and particle pairs can be created from and decay into each other.

The CAGI and natural selection have produced some very complex pieces of matter with some very interesting emergent qualities. While we dig deeper into the secrets of nature, we should always be ready to take a step back and recognize such qualities. It is easy to miss the forest through the trees, in this case the emergent qualities through the constituent parts.

Our creation of a new vision of the early universe continues with some speculations on some unsolved cosmic mysteries in the next chapter.

Chapter Seven
Speculations on Unsolved Mysteries

We have nearly completed our construction of a new vision of the early universe. We must yet examine what that new vision is telling us about *reality*. First, however, there are a few remaining mysteries surrounding aspects of the universe worthy of addressing. Unlike the resolutions of mysteries expressed thus far, my answers to these mysteries have little observational support. They do, however, make a good deal of sense given our premise that the universe started from an expanding speck of pure energy that naturally evolved. We start with one of the biggest mysteries in cosmology today – what is *dark matter*?

Speculations on the Origin of Dark Matter

Astronomer Vera Rubin while studying the rotation curves of several galaxies noticed that they were not rotating at the speed they should have given the apparent amount of mass they possessed. It has since been theorized that the added gravity influencing the rotation of those galaxies is due to an unseen quantity of mass in a halo around the galaxies, though the source of the mass is unknown. Let's speculate on what that mass may be.

One of the consequences of the evolutionary process is the creation of a great deal of waste and debris from the relatively short lives of many evolutionary experiments. On earth, debris is broken down in our oceans and soil. In space, unused particle matter is broken down by annihilation and collision, and that debris is left to drift with the cosmic tides.

Using Tors, Tor-Chains, and Tryk combinations, the evolutionary process could build all kinds and shapes of particles and anti-particles. Stable, long-term creations, however, are always limited by the hazards of the environmental frenzy and the competition for survival inherent in natural selection. As is often the case with natural selection, some of the stable creations that do not go on to develop further might continue to exist but remain developmentally stagnant. Accordingly, our Penergy medium will likely contain odd particles and particle fragments that were stable enough to withstand the frenzy, but not flexible enough to become a part of something larger and more complex.

Those odd Tors, Tryks, Chains, and particle fragments left behind in evolutionary development are unseen by us because only particles with electric charge can absorb/emit photons. Any particle remnant with an attractive or repulsive charge would float around, bumping into other fragments until it combined with the right fragment(s) to neutralize its charge and be unable to absorb/emit photons. Accordingly, one might call these now neutral fragments collectively *dark matter*.

Physicists Paul Davies and John Gribbin expressed a similar idea in their book, *The Matter Myth*. On the origin of dark matter, they said, "Nobody is sure what the invisible stuff is, but the best bet is that it is an unseen residue of exotic subatomic particles left over from the big bang."[1] Some theorize dark matter to be some exotic new particle. It seems more reasonable for dark matter to be normal particle fragments, however, since it is observed that the density of dark matter has the same magnitude as the density of visible matter.[2]

These dark matter fragments would have floated around bumping into each other, gradually expelling their kinetic energy. No longer capable of combining or absorbing more energy, they were left to gather in clumps due to their own gravity. Scientists believe those clumps represent five-times more dark matter than the visible matter in the universe, and that its cumulative gravitational influence played a part in the shaping and distribution of the developing galaxies.

Much of the left-over particle debris would likely be in the form of very lengthy left and right-spinning Tor and Tryk-Chains. Consequently, as the cosmic turmoil around blackholes occasionally breaks up some of the dark matter, exposing those long chains, we would expect to see an unusual number of magnetic fields in their vicinity, which have been observed.

Dark matter fragments possessing little energy and no charge will be difficult to detect. However, if we could isolate a portion of dark matter and spray it with a high energy laser, we might break up the connections, re-establishing its altered configuration as charged particles. This would allow it to once again absorb/emit photons, giving us the opportunity to see it. But, being comprised of the same Tors and Tryks that compose matter in general, I wonder if we will be able to recognize it.

Dark matter is the *boneyard* of particle fragments but collectively it has substantial mass. As mentioned, it is believed to be adding gravitational influence to our galaxies, which explains the unusual rotational speeds of some of them. There may not be as much dark matter in the universe as scientists calculate. Some of that added gravitational influence affecting galaxies could be coming from the SMBH at their center in the way of *dynamic gravity,* discussed next.

Speculations on the Origin of Dynamic Gravity

The strength of a blackhole's gravitational field is believed to be proportional to its mass, and its mass is calculated based on its gravitational effect on surrounding stars and gases. In other words, our calculation of a blackhole's mass is dependent on Einstein's equations used to calculate its gravitational field.

In the Tor Model, the measure of a blackhole's mass and gravitational field is proportional to the density of its Penergy, which in turn has a relationship to its spin rate. This is quite a departure from current theory, but perhaps such a departure is necessary. As Astrophysicist Stuart Clark points out, "...general relativity is the ultimate gravitational theory. Yet it can't explain black holes. If we believe that the universe is understandable through mathematics, the black holes are telling us that there must be a deeper theory of gravity to be found."[3]

According to Einstein, mass couples to spacetime,[4] and as previously observed massive spinning objects such as blackholes are known to *drag* the surrounding spacetime (*Penergy medium*). Perhaps it is the spinning of the blackhole dragging the Penergy medium that has a significant influence on, if not creating entirely, the gravitational field. Spin being responsible for dragging the Penergy medium is accomplished in the same way Tor spin creates a gravitational field around smaller massive objects, as discussed in Chapter Four. It is a part of the *relationship* the Penergy has with its prodigy - Blackholes and Tors.

Prediction: Someday a sharp young astrophysicist will figure out a way to measure the gravitational field surrounding a SMBH and discover that the strongest area of its gravitational field is at its equator, where it is spinning the fastest. This will confirm that it is not simply the mass of the SMBH that is creating its gravitational field, but the *spinning* of the SMBH.

As mentioned in Chapter One, it makes sense that matter inside a blackhole goes through transitional phases as it becomes denser. The blackhole crushes matter back into Penergy, the Penergy into denser and denser Penergy, and finally into its NIB (*Near Infinite But* not quite) state. Stephen Hawking told us that once matter passes the event horizon that, "... the matter inside the blackhole would be trapped and would collapse into some unknown state of very high density."[5] This is consistent with the Tor Model's vision of the phase transitions inside a blackhole.

In the Tor Model, *gravity* can only crush matter down to a certain level - when it is no longer an independently spinning entity and turns into a glob of fluid-like Penergy. This probably happens just inside the fringe of the blackhole. Beyond that point the further condensing is not due to gravity as we know it, but due to the internal dynamics of the fast-spinning Penergy itself.

The heart of the blackhole is a long distance from, and no longer influenced by, the gravitational curvature of the surrounding thin Penergy medium where Einstein's equations work quite well. Inside the blackhole there may still be an influence pulling the Penergy toward the center, but the fate of the Penergy being condensed is now determined not by what we consider as gravity but by the dynamics inherent in the internal Penergy condensing process, taking it into denser and denser phases. That state is beyond the realm of our current idea of gravity described by general relativity, so it will require further theory and perhaps new mathematics before it is understood.

The radius to which a mass must be compressed down to form a blackhole is called the object's Schwarzschild radius, named after Karl Schwarzschild, who was first to find a solution to Einstein's equations describing a spherical blackhole.[6] The bigger the mass of the blackhole, the larger the

Schwarzschild radius. In the Tor Model, this proportionality eventually changes.

The effect of the slow building NIB-state at the heart of a blackhole is called *dynamic gravity*. It is supplemental to that of normal gravity, which is the curvature of the Penergy medium caused by the contact with the spinning blackhole. Dynamic gravity may not manifest itself until the blackhole is quite large and has processed a tremendous amount of matter and Penergy. The tell-tale sign of a NIB-state would be when the blackhole's spin rate goes up and its event horizon begins to shrink. Some SMBHs may be in the initial phase of that state already, as spin rates have been measured to be fifty to ninety-five percent of light speed.[7]

This phenomenon of added mass causing the diameter to shrink has been observed in neutron stars where the energy density is quite high. In contrasting the customary effects of gravity to neutron stars, Astronomer Luciano Rezzolla tells us that, '...the opposite is true, as the mass grows, the radius tends to decrease."[8]

The spin-rate/circumference relationship is subject to the same dynamics an ice skater employs to spin faster by pulling in her arms, shrinking the radius of her mass, driving up the angular momentum and spin rate. In the blackhole's case, it is the faster spin rate that is causing the radius of the mass to shrink. Theoretically, if we fed all the matter and Penergy in the universe into a single, giant blackhole, its spin rate would continue to rise, its circumference would continue to shrink, and eventually it would return to the speck of NIB-density Penergy from which our universe began.

Scientists have put forth theories that support stronger gravity and less reliance on dark matter to explain unusual galaxy rotation. Israeli Physicist Mordehai Milgrom proposed simple changes to Newton's laws that when applied to galaxies turned out to match their rotation speeds quite well. His theory became known as MOND for Modified Newtonian Dynamics. MOND theory is out there but is not yet well accepted. It did, however, show that the mass of a galaxy is related to its rotational velocity.[9] Perhaps the final theory affecting galaxy rotation will be something in between MOND, dark matter, and Dynamic Gravity. There is no evidence yet for dynamic gravity, it is only speculation. But someday an enterprising young physicist may develop the mathematics to support such a theory.

I love blackholes. They offer an endless variety of topics about which to speculate. I've always doubted the idea of supermassive blackholes formed from the collapse of primal gases or the merger of multiple galaxies. The idea of SMBHs forming out of dense pure energy is much more appealing, but that notion raises some questions, such as how a SMBH made of pure energy differs from the speck of pure energy from which our universe began. This got me thinking about the presence of gravity without mass and how a super-giant SMBH could ever be shrunk down to the size of a speck of pure energy. From this line of thinking I developed the notion of *dynamic gravity*. There is no evidence for such a phenomenon, but it is an interesting speculation.

Another important aspect of particles is the way they are created and decay. Their creation might be related to the presence of dark matter, so let's continue our speculations and look at these interesting phenomena.

Speculations on Particle Creation & Decay

Particle Creation

Dark matter particle fragments could remain very useful. After the birth of the universe, once the density of the Penergy drew thin enough, it spun Tor type particles at will. That particle producing density has long since passed. Now it is only when there is a knot of Penergy of sufficient density that it will produce particles. The particles it creates may well depend on the *source* of the knot of Penergy.

If the knot of Penergy was created in the collision of two high-energy particles, such as two protons traveling at near light speed in a collider, then the knot of Penergy is more likely to create one or more large particles that instantaneously resolve into a flood of lighter particles. That phenomenon is observed daily in particle colliders and is how we confirmed the existence of the first and second generation of particles, as well as a host of other heavy unstable particles.

The source of the knot of Penergy could also come from an existing single, high-energy particle, such as a gamma ray. Gamma rays are high energy photons created in explosions, blackhole jets, and other high energy events occurring in the universe. If the knot of Penergy was created from the natural decay of a gamma-ray photon, the Penergy knot will usually spin itself into a particle pair, as described in Chapter Two. It has the potential to create any particle pair commensurate with its energy content. But how does that knot of Penergy know what to create, or how to create it? We can speculate as to how the knot of photon Penergy will respond as it changes into a particle pair.

Perhaps this knot of Penergy would not necessarily spin *fresh* particles. Instead, as the Penergy knot begins to create a pair of left and right-spinning particles, it picks up Tors, Tryks, photon fragments, or other fragments from the *boneyard* and uses them either directly, or indirectly as a blueprint, to construct the largest particles possible given the available Penergy.

Scientists have observed a high-energy photon create a matter and anti-matter pair of protons. According to the Standard Model protons are complex particles made up of two kinds of quarks and a sea of eight different gluons. The anti-proton is just as complex, made up of the same particles with the opposite charge. If two complex matter/anti-matter particles are going to be created out of a knot of Penergy, it seems more plausible they were rapidly assembled directly from the boneyard of particle fragments than having been created anew with such complexity simply out of the Penergy alone without such a blueprint. It seems to me to be an extension of the evolutionary process of combining existing building blocks to create complexity. There is no evidence for such a particle creation process, but it is an interesting speculation.

Particle Decay

Particles decay (*decompose*) in two ways, *forced decay* and *natural decay*. Forced decay occurs when a particle cannot subsist in the existing density level of the Penergy medium. This occurs in an accelerator or collider when a knot of Penergy creates a very large particle. The large particle immediately decays into smaller particles, and in some cases those smaller particles resolve into even smaller particles, until all particles produced can reside in the existing Penergy density.

Natural decay occurs when a large atom such as plutonium changes by ejecting an alpha particle (*two protons and two neutrons*) from its nucleus, which is a process known as *Alpha Decay*. A down quark decays naturally by ejecting an electron and neutrino, which is known as *Beta Decay*. The cause of this particle instability is not well understood. QFT associates the decay with the *weak force* that encompasses the sudden creation of a very large virtual boson.

The Tor Model associates decay by other means. It could be that in the case of Alpha Decay, the jiggling of a large nucleus occasionally distorts its shape allowing the repulsive forces between protons to overcome the short-ranged nuclear force, enabling the alpha particle to escape. Chemical Physicist

Michael Munowitz reminds us that despite the strong nuclear force, the electric repulsive force between protons is always present and the potential for disaster never disappears. When the repulsive force overtakes the strong force, radioactive decay results. The nucleus can spew out a proton, neutron, or alpha particle. After the decay, all the protons and neutrons are accounted for. The particles are shuffled about but are otherwise unchanged.[10]

In the case of Beta Decay, most beta particles are ejected at speeds approaching that of light.[11] This suggests that the decay is triggered by the absorption of a high energy photon that overcomes the binding energy of the electron connection, breaking that connection and allowing the electron/neutrino duo to escape from the down quark. In the Standard Model, the escaping neutrino is deemed an anti-neutrino to balance the negative charge of the electron in support of charge conservation. The Tor Model sees the neutrino as a neutral particle. Since it is comprised of composite particles, specifying its charge is unnecessary.

Standard Model physicists would argue that the created particles from the beta decay did not come directly from the original particle, but from a virtual boson that mediated the decay. In the case of the decay of a down quark, the boson in question would be a virtual W- particle. It means that the less than 5 MeV sized down quark suddenly emitted an 80,000 MeV virtual W- that then immediately changed back into a less than 5 MeV up quark, plus an electron and neutrino. Assuming that is even possible, it seems like a laborious way for nature to work.

Heavy particles like the W boson have been observed in colliders and physicists have developed the math to support such an inefficient interaction, but relying on heavy, virtual particles to explain particle decay seems dubious. It is also unnecessary if our Sub-A's are in fact composite particles. A more reasonable explanation for the result of the down-quark decay is again that the electron and neutrino were present within the parent down quark before the decay. Admittedly, there is no direct evidence for such a composite configuration, but we just don't have the technology to break down those tiny Sub-A's to that degree.

As detailed earlier, physicists theorize that after each interaction, which includes a decay, particles are not simply reconfigured but instead all new particles are created.[12] This seems rather extreme, and it makes more sense for nature to handle particle decay like it does atom decay. When an atom decays by spitting out an alpha particle from its nucleus, it simply leaves behind an atom with different atomic numbers and characteristics. The protons and neutrons are reshuffled but otherwise remain unchanged. Particle decay seems more about rearranging existing components than creating all new components from scratch. Nature would work far more efficiently that way.

While we are speculating, let's look at another interesting topic about the universe – special relativity. Einstein gave us the equations showing how space and time are mathematically relative to each other, but the equations do not explain *why* this is so. Let's examine the relationship between space and time. They are related, but what is the source of that relativity? You will find this speculation a fun venture.

Speculations on Why Space & Time are Relative

The Penergy medium is a wonder to behold. Like the tough fabric it is, it can withstand the contractions, bending, and warping by massive blackholes, and at the same time act as stretched cellophane on which subtle vibrations can be transmitted allowing tiny particles to communicate. It can withstand the comparatively large gravitational waves created by crashing neutron stars, while at the same time holding a single particle's delicate field. It can hold the trillions of individual particles, their vibrational signals, all the photons of radiation and their related electromagnetic fields, and simultaneously allow the tiny signal from my friend's cell phone in Paris, via satellite, to find my cell phone in California, and relay her voice clearly and distinctly. The Penergy medium is truly fantastic and dynamic. It is also closely related to time and may be responsible for why space and time are relative to each other.

We can't see or feel space or time, but they remain very real to us. We derive their existence by sensations of objects and the movements of those objects around us. We once thought of space and time as distinct entities but have since learned from Albert Einstein that they are connected. Let's examine how that could be.

A Thought Experiment

Let's do a thought experiment. Imagine a capital T with a railroad track running across its top and a narrow conveyor belt running down its base. A mailman on a passing train wishes to toss a bag of mail onto the conveyor belt. The mail bag resting on the moving train has momentum, which will carry it forward as it falls toward the conveyor belt. Knowing the speed of the train and the drop distance to the conveyer belt and using Newton's laws of motion one can calculate when in advance to let go of the mail bag to hit the conveyer belt.

Now imagine the same setup with the capital T, but this time from the train we are going to emit a photon down the center of the conveyor belt. For this to occur, where should the train be when the photon is emitted? From all that I have read, the emission point on the train should be directly adjacent to the center of the conveyor belt. The photon will not travel forward in the direction of the train's motion as the mail bag did. When the photon comes into existence, it immediately steps out above the conveyor belt and shoots away in a *straight path* at light speed, and it will stay on that straight path until influenced by a gravitational field or some form of matter.

This makes sense because whereas the mail bag was *in existence* and subject to the laws of physics regarding the movement of it and the train, the photon as we know it *did not exist* in the same sense until it was emitted from the train.[13] Further, the photon was separate from the train the moment it came into existence, so the train had no influence on the photon's movement.

The photon was only subject to the laws of physics regarding the Penergy medium which, is the only realm in which a photon exists. This goes along with the idea that the speed of light (*the photon*) is always measured the same, whether emitted from a source traveling in the direction of emission or away from the direction of emission. The speed and direction of the source of emission is immaterial to the photon.

Proof that photons travel only in straight lines irrespective of the movement of their source is found in experiments measuring the distance to the moon, where photons are shot to reflective mirrors left on the moon's surface. According to the experimenters, a light beam from earth that strikes the mirrors is reflected back on the exact same path that the beam took to reach the mirrors.[14]

If the light beam shot from a spinning earth is reflected on the same path from the non-spinning moon, then obviously the spin of the earth had no effect on the path of the beam. It is a straight shot in both directions. As physicist Brian Clegg says, "...light always goes in a straight line. Period. There is no arguing with this."[15]

Another Thought Experiment

The conclusion drawn from the above thought experiment is that a light beam may bend due to gravity, but it will not bend due to the movement of the source of the beam. If this is true, it brings into question an often-used argument for time dilation -- the light beam bouncing within a moving spaceship scenario.

The argument is that two mirrors perfectly parallel to each other are set up on the floor and ceiling of a spaceship with a light gun mounted at the upper mirror aimed at the lower mirror. Once the spaceship is under way at high speed the light gun fires light (*photons*) at the bottom mirror. The light hits the bottom mirror and is reflected up, hitting the top mirror and reflected back down, etc.

From inside the spacecraft the photons are simply seen bouncing up and down between the mirrors. For an observer outside the spaceship, however, she sees the same light moving in the direction of the spaceship making a series of W's between the mirrors. Given the light is moving in a W, a distance much greater than the direct distance between the mirrors, the photons must be traveling a much longer distance from mirror to mirror. Because light only travels at one speed, the only way for it to travel the longer distance is for *time to slow* within the fast-moving spaceship.[16]

At first impression one may wonder how the photons traveling a lateral distance could be said to be traveling a greater distance necessitating more time than it takes to bounce them up and down between the mirrors. The time it takes for a bullet dropped from five feet to hit the ground is the same for a bullet fired from a five-foot-high level rifle, even though the bullet may travel a thousand yards laterally before striking the ground. Lateral movement does not add any time to falling bullets, why should it for bouncing photons?

That argument aside, let's continue with our thought experiment and say we make the bottom mirror very tiny, perhaps only a few atoms wide. It is still parallel to the top mirror and still capable of reflecting the light gun photons.

Now we get the spaceship up to speed and fire the light gun. The light travels toward the bottom mirror, but this time it does not reflect up to the top mirror. It missed hitting the tiny bottom mirror because the movement of the spaceship moved the bottom mirror out of the way before the light reached it.

The light when emitted from the light gun was set on a *straight path* downward, and but for the high-speed spaceship moving the bottom mirror out of the way it would have struck the mirror and been reflected. Having missed the bottom mirror, the light does not move in a W or travel a greater distance by being emitted from inside a moving spaceship. For it to be reflected, the light would have to move in an arc back toward the bottom mirror. If that were the case, light would bend in an arc every time it was emitted from the spinning earth, but as discussed above, apparently that is not the case.

Relativity

The above analysis seems correct, but whether valid or not it warrants examining the relativity between space and time a little closer. According to Einstein, the gravitational field of a large mass will cause time to slow. Time is relative to gravity as well as space due to the equivalency between a gravitational field and acceleration. Perhaps time is not relative to space directly, but relative to something equivalent to space that includes gravity. Let's explore what that could be.

Space is the void between stars and galaxies. We call it the vacuum of space, but we now know there is no such vacuum. Space is permeated with dark energy, which we have identified as Penergy. So, when we are saying 'space', we are really referring to our Penergy medium.

For Penergy and time to be connected and relative to each other, it would mean that time is somehow a dynamic of Penergy, perhaps a function of its *density*. Time is relative to something changing, and it may well be that Penergy density affects how quickly any change takes place. That may sound crazy, but let's examine the notion and see where it takes us.

If time were a function of Penergy density, then a change in density would mean a change in the passage of time. As we have seen, Penergy was once very dense, but it has been thinning since the universe began. Its density now only changes to a higher density when it is condensed or compressed. Let's examine how those conditions could come about and whether they are related to a change in time. Einstein identified two situations causing time to change: *when near a mass with a strong gravitational field*, and *when flying through space at a very high rate of speed*.

Time Dilation Due to Gravitational Condensing

Einstein's general theory of relativity taught us that the gravitational field of a mass affects both space and time, and that the greater the gravitational field the more space is curved, and the slower time passes.

One of the proofs of Einstein's theory is that the GPS satellites orbiting the earth must adjust for the time differential between the earth's surface and their orbital height above the surface, otherwise the system would be thrown off significantly within a short time. This is due to the clocks in the satellites experiencing a lesser gravitational-field influence than the clocks on the surface of the earth. Time is running slower on the surface than it is in orbiting satellites twenty thousand kilometers above the earth's surface.

As we have established, a body with mass curves the geometry of the surrounding spacetime creating what appears to us as an attractive influence. That attraction would likely not only affect other masses, but the surrounding Penergy medium as well. It too would be drawn into the gravitational field, condensing at the surface of the mass. The condensing of the surrounding Penergy would cause it to have a higher density in proportion to the mass's gravitational field. That higher density of Penergy could be what is causing time to slow, not simply the gravitational field produced by the mass.

In the case of a blackhole, the strength of the gravitational field is enormous, having a significant influence on time passage. Theoretically the crew on a spaceship heading into a blackhole would pass through the event horizon without noticing any time change. A person watching the spaceship from a distance, however, would see time in reference to the craft gradually slowing and seemingly come to a stop as the spacecraft reaches the event horizon.[17] That notion has obviously not been validated, but the point is that the stronger the gravitational field, the slower time passes.

The Penergy medium does not have mass as we define it. It's fluid-like consistency makes it respond differently to a blackhole's gravitational field than does mass. While mass is drawn into and consumed by the blackhole, Penergy is drawn close but its high rate of spin around the blackhole gives it angular momentum, which pushes the Penergy medium away at the event horizon. If the blackhole is rapidly spinning, as most do, the Penergy is pulled in tightly to the blackhole, but is held at bay. If fast spinning blackholes eat Penergy like they do objects with mass, the first blackholes would have likely consumed all the remaining Penergy in the universe and we would not be here.

The fast-spinning blackholes drag the surrounding Penergy with it, performing the *frame dragging* described earlier. The surrounding Penergy builds in density as it is pulled close, causing time to slow as seen by the observer watching the rocket approaching the event horizon.

The much higher Penergy density at the event horizon likely also causes particle pairs to be created consistent with the strength of the density. Those particle pairs include the first and second generation of particles – the heavier quarks and leptons. Some of those particle pairs would be drawn into the blackhole, and some would escape, being slung out into space. This could be why the earth is constantly bombarded by cosmic rays – high energy particles and anti-particles from space having an indeterminant origin.

Time Dilation Due to High-Speed Compression

The other circumstance in which time is dilated according to Einstein is when an object such as a rocket is traveling rapidly through space. The closer the rocket is traveling to the speed of light; the slower time passes for those within the rocket. According to Einstein, a mass such as a rocket traveling through space [*the Penergy medium*] would grow in mass and shrink in the direction of travel the faster it moved.[18]

The high-speed mass moving through the Penergy medium would cause the Penergy around it to compress. The compression of the Penergy would raise its density in proportion to the mass's speed and growth in its own density. The change in the spacecraft's density would be enormous. A mere particle traveling at .9999% light speed would have a density 10,000 times greater than its rest mass.[19] Consequently, the greater density in mass would compress and raise the surrounding Penergy density, slowing time for those inside the mass (*rocket*). The same effect would even apply to something as small as a particle. A particle speeding through our ubiquitous Penergy medium will also compress the medium, raising the density around it and slowing time for the particle.

Einstein pulled together space and time into one concept. We can see now why that works; because space is made up of Penergy, and Penergy density affects particle movements. Penergy density doesn't change "time" per se, only the movement of particles, and that rate of change in position (*think atomic clocks*) is how we measure time. So, Penergy is responsible for both *space* and *time*, making Penergy and Spacetime the same thing.

There is no hard evidence for either of these time dilation theories being related to Penergy density. Curiously, however, in a July 2023 article in *Live Science*, Ben Turner reports that scientists have found new evidence for time having passed much slower in the early universe. This arose out of a twenty-year study of twelve-billion-year-old quasars and the analysis of the different wavelengths they emitted. The findings are said to have finally proven a prediction made by Albert Einstein.[20] The Tor Model posits the Penergy density being much higher in the early universe, so there may be a correlation with time passing slower. Notwithstanding this scant evidence, the relationship between Penergy density and time dilation represents a possible nexus between time, spin, gravity, and space. If proven valid someday, it would mean that Penergy truly is *spacetime*.

As the Penergy density thinned, the rate of change in movement of all things would have grown faster. If that faster rate of change would in turn make the Penergy thin faster, then the effect of one on the other would cause the expansion of the universe to gradually accelerate. With nothing to oppose the expansion initially it would have expanded quickly, but only temporarily. Once the SMBHs began exhibiting a gravitational influence it would have slowed that expansion, but again only temporarily, perhaps only several billion years. Once the universe grew to the extent the galaxy/gravitational density fell below a certain value, the expansion would again overtake the gravitational influence and begin accelerating again, which apparently it is doing currently. That acceleration may not be permanent, however, as we do not yet know the effects of dynamic gravity on the universe. My sense is that it will once again eventually slow, if not stop, the acceleration, and possibly the expansion all together.

Einstein's work on Special and General Relativity was truly genius. It should be noted that in both instances he developed the concepts before developing the math to support them. In the case of General Relativity, it took years to develop the math and he had help from many colleagues. But the final product was and is a true testament to how mathematics can be used to describe the laws of nature.

The final equations were so concise and elegant they caused Einstein to change his style for developing concepts to describe nature, putting mathematics first. He worked for years trying to discover how the forces of nature could be unified using a mathematical approach. He never succeeded in the endeavor. I often wonder whether he would have been more successful had he continued first using his imagination and thought experiments, rather than mathematics, as he had in the development of relativity.

Mathematics is unquestionably an excellent tool, but it shouldn't be relied upon solely to make future discoveries about nature, though that seems to be the trend. I believe there is still a virtue in placing the conceptual imagination before the mathematical imagination, and to do otherwise runs the risk of a loss of both substance and clarity. Astrophysicist David Lindley tells us, "Traditionally, the greatest theorists of bygone days have begun by reasoning about the conceptual physics of a problem, pondering mechanisms and ideas of cause and effect, and have then worked their way, often with great difficulty and many wrong turns, to a correct mathematical description."[21] To the ears of this old philosopher, that sounds to be a much better approach to resolving the mysteries of our universe.

Since mathematics plays such a big part in the theoretical developments and description of the workings of the universe, we should get a perspective on its evolution and value as well. But first, some more speculation.

Our Future in Space

Regarding particle production due to Penergy compression, I visualize that someday when we have the technology to travel very quickly through space, our spaceship will compress the Penergy at its leading edge to the extent the raised Penergy density will create particles. Given we possess the technology allowing the spacecraft to absorb those particles, we will be producing our own source of raw material from which to fuel our existence in space.

Mathematics and Uncertainty

Sir Isaac Newton is credited with creating/organizing the laws describing mathematically the motion of things and the way gravity plays a part in their movement, necessitating the development of Calculus. The laws and math relating to Newton's work are also referred to as classical physics. The big advancements in Newtonian math and physics came in 1865 with Maxwell's equations on electromagnetism and in 1915 with Einstein's equations on general relativity. Both were describing the details of the *fields* we know to exist and the *forces* they incorporate, together comprising the heart of physics.

Maxwell had to use a new type of math called vector calculus, which was necessary to show how changes take place within electric and magnetic fields in a three-dimensional space. Einstein had to use a new type of math called tensor calculus, which was necessary to show how changes take place within a gravitational field in a four-dimensional space. Both men built upon the work of others, but the concepts and mathematics that finally evolved are nonetheless considered inspirations of genius. These two advancements reinforced the idea that mathematics was capable of accurately describing the laws of nature.

Maxwell and Einstein's work gave us a new and odd perception of reality, one not easily believed for some time. Their work laid the foundation for further discoveries in the subatomic arena where scientists were exploring matter at smaller and smaller scales, until they could no longer measure the location and momentum of a particle simultaneously. Realizing experimental results could only be predicted as a *probability* of an outcome, an entire new mathematical formulation called Quantum Mechanics (QM) was developed.

Though QM gave only probabilities, it nonetheless proved to be quite accurate in describing the very small (*quantum*) aspects of the universe. Its development caused our picture of reality to be distorted even further. Theorizing particles to be waves without a precise location, coupled with an inability to measure their location and momentum simultaneously, meant that knowing a particle's momentum also meant that its location could be *anywhere*. Our picture of reality had become truly weird, which is where we are today. Perhaps it is not quite as weird as believed, however, as we will soon see in our examination of *reality*.

The Tor Model should easily adopt the Standard Model's mathematics regarding the measurements and relationships between particles, fields, and forces. Again, it is not the core mathematics that the Tor Model takes issue with, but the *interpretation* of some of the mathematics. The Tor Model may not necessitate a new math, but once its perspective on the evolution of the universe is evaluated and tested, physicists may have to create new terms and values for Penergy, Penergy density, the interior of a blackhole, and the multiple aspects of a particle's personal field.

We have come to rely on mathematics more and more, perhaps to our peril in understanding some aspects of the universe. Like any language, no matter how well formed and concisely stated, its content is subject to interpretation.

Consequently, though math has played a significant role in our understanding of the universe, it cannot always be relied upon to describe the exact picture of nature, though occasionally it gives us glimpses of aspects we would not have otherwise recognized or even entertained.

Math is wonderful but perhaps we are giving it a greater emphasis than it deserves. It is a tool created by humans that consists of symbols representing measurements and concepts, and rules to manipulate the symbols. Its success in accurately describing nature in terms of the relationship between those concepts speaks volumes for the creative genius of our species. Visiting aliens will no doubt have their own mathematics consisting of symbols, concepts, and rules of manipulation, but it won't be a ten based numerical system of numbers and symbols divided by an equal sign. That is entirely a tool created by humans that likely exists nowhere else in the universe. Our concept of mathematics is our own, not a universal feature.

As Physicist Savine Hossenfelder tells us in her recent book, *Existential Physics*, "Physicists may not consciously subscribe to the idea that math is real and when asked will deny it, but in practice they do not distinguish the two. This conflation has consequences, for they sometimes erroneously come to think their math reveals more about reality than it possibly can."[22]

* * *

We have built an early universe through observations and deductions that blend smoothly into the universe believed to have existed 380,000 years after the Big Bang. The observational evidence for the Big Bang theory mentioned in the Introduction is still intact. What makes our reconstruction of the early universe compelling is that we did it without anything exploding, popping into existence, or inflating. There was quite a bit involved, but every effect we observed was a consequence of a natural and reasonable cause; a sequence one might call *natural cosmic evolution.*

With our reconstruction project complete, we need now only look at what that construction process has been about. What does it tell us about the bigger picture of the universe? What is it telling us about *reality*?

Chapter Eight
The Origin of Reality

We have created a sound theoretical foundation to our early universe that accounts for the creation of SMBHs, the Cosmic Web, particles, fields, forces, and possible subatomic particle configurations. We may never be able to prove the existence of Penergy, pure energy blackholes, or early elementary particles, but their theoretical existence answers many questions and seems quite reasonable given the observational evidence.

We shall now evaluate whether we are appropriately interpreting this new vision. Stepping back and taking in the breadth of our project we see billions of spinning, pure-energy, supermassive blackholes that were initially surrounded by very tiny, early elementary particles. Those particles have combined and grown in Levels, building in complexity. Revealed at each Level were previously unseen *emergent qualities* that have included wondrous attributes such as fields, forces, photon absorption, and the variety of atomic combinations that will be used to build our universe from here.

That is what we see on the surface, but is there more going on beneath the surface? Can the presence, behavior, and interaction between all these components give us a picture of a deeper *reality*? Can our Standard Models and the equations of general relativity and quantum theory be the path to discovering a deeper reality? According to physicist Lee Smolin, physics should be more than a set of formulas that predict what we will observe in an experiment; it should give us a picture of what reality *is*.[1] The discoveries and theories of particle physics and cosmology have given us road signs pointing to a subsurface reality. Let's head down that road to reality and see where it takes us.

Finding Reality

Let's start with a caveat: we are limiting our definition of reality to what is happening in the universe independent of a life-form or conscious observer. This allows us to sidestep all the philosophical issues regarding the fallacies of our perceptions, limits to our sense organs, etc. The reality we seek is the one that is happening independent of us.

For physicists and cosmologists, reality is ultimately expressed in equations. So far, matter, energy, fields, forces, their movements, and relationships have lent themselves to being described quite accurately in mathematical terms. That success says a great deal about human curiosity and ingenuity. We seek to understand our world. We peer into the depths of space and the Levels of matter in hopes of learning the laws of nature that will explain who we are, where we have been, and where we are going. We are a part of a mysterious nature, but one that can apparently be understood. There is much we know about it, but much we don't know. Given the success of our understanding so far and the success of our describing it mathematically, many believe there must be a description of a single reality that encompasses all there is.

The Standard Model of Particle Physics and the Standard Model of Cosmology are touted as being two of the greatest accomplishments of the twentieth century. Given what they encompass and what they have accomplished, they are indeed monuments to human endeavor. The mathematics supporting these Standard Models isn't full proof, however. Physicist Frank Wilczek tells us the Standard Model of Particle Physics is flawed; its equations are lopsided and contain several loosely connected pieces.[2] These models have provided us with a reasonable picture of the universe and how it works, but there are many missing pieces to the picture (*dark energy and dark matter*), and some areas are vague and a challenge to credulity (*inflation and missing anti-matter mystery*). These two models, as wonderful as they are, cannot provide us with a complete and cohesive picture of the universe, and by themselves are unlikely sources for discovering an underlying reality.

General Relativity (GR) does a wonderful job of describing space, time, gravity, and the larger aspects of the universe. It predicted blackholes, gravitational waves, and time dilation, all of which have been confirmed many times. Its equations, however, only address those larger aspects of the universe and cannot be used to describe the micro-world of quantum particles. By itself it is not likely the source for a complete picture of reality, but its success certainly suggests it will have to be accounted for in such a vision.

The mathematics of Quantum Theory (QT) accurately describes particles, fields, and the smallest aspects of the universe, but there is much controversy over whether QT represents an underlying reality. Physicist Niels Bohr and other architects of quantum mechanics believed it to be a good tool for describing/measuring aspects of particles, but they did not believe it should be taken as a broader theory as to what is real. Einstein was uncomfortable with quantum theory, and over the years many others have felt discomfort in the disparity between the world as we see it and the accepted model of the quantum world where nothing is really what it seems. Physicist Michio Kaku says, 'Quantum Theory is the most ridiculous theory ever proposed in the history of science.'[3] But, as he points out, it works and has withstood every experimental test hurled at it. Still, it is difficult to believe that a theory believed to be the most ridiculous theory ever proposed would at the same time form a sound basis for describing reality. QT accurately describes our micro world mathematically, but its interpretation of that world leads to some unusual conclusions, often referred to as quantum weirdness. Can such weirdness serve as a basis for our reality? We need to examine Quantum Theory a little more closely.

Quantum Theory

The development of QT is an interesting story that starts well before Isaac Newton when fundamental matter was theorized to be made up of tiny, physical objects called atoms. In the seventeenth century Robert Hooke and Christian Huygens developed a wave theory of light, while Newton had developed a 'corpuscular' theory of light, both of which contained reasonable evidence for their respective theories. Later developments confirmed light to theoretically have both wave and corpuscular characteristics.

With Einstein's 1905 paper on the photoelectric effect, light energy was proven to be particles (*photons*), thus giving them definite particle-like *and* wave-like characteristics. Physicist Louis de Broglie later advanced a theory that *all* particles have wave-like attributes, which proved accurate. The recognition that two related particle attributes could not be precisely measured simultaneously gave rise to the Heisenberg Uncertainty Principle (*discussed shortly*). That in turn gave rise to the concept of superposition, wherein a quantum object could be in two states at the same time.

All these observations and developments resulted in the need for a new mathematical structure that recognized that experimental outcomes involving quantum particles were not certain and subject only to probability. With the development of Schrodinger's equation containing a *wave-function* to represent the wave-like potential and probability for particle measurements, Quantum Mechanics was born. All this brilliant work was accomplished before 1930. It has been refined and extrapolated since then, for the most part making the math better, but its interpretation even weirder.

There were different schools of thought on how to interpret quantum mechanics. One school was to make no interpretation at all: *the mathematics of quantum mechanics works very well so don't bother with interpreting its meaning.* That school of thought was known as the Copenhagen Interpretation, which was apparently influenced by Logical Positivism, a popular philosophy at the time. That philosophy was founded on the idea that the only things we can know is what we observe, and what we are not observing does not exist, or at least asking questions about it is senseless. That particles don't exist if we are not observing them, and don't have attributes until we measure them, is called a non-realist philosophy.

According to the Copenhagen group, quantum mechanics tells us nothing whatsoever about the world. It is merely a tool for calculating the probabilities of various outcomes of experiments. Due to the strong character of its primary proponent, Swedish Physicist Niels Bohr, that interpretation dominated for many years and is still taught in universities today. One of the reasons it has sustained over the years is that there is no consensus on a better interpretation of Quantum Mechanics.

Einstein was a realist. He believed particles known to exist continue to exist whether observed or not, and that particle attributes exist both before and after measurement. For many years he carried on verbal battles with Niels Bohr who of course believed that any discussion about unobserved particles and their attributes was senseless. Bohr was perfectly content with both quantum duality and quantum uncertainty and did not believe an underlying meaning was necessary. His philosophy was summarized with the expression, 'Shut up and calculate.'

An attempt to put quantum mechanics back into the 'realist' philosophy camp was put forth by Louis de Broglie, and later by Physicist David Bohm, called the *pilot-wave theory* discussed in Chapter Three, but it was never accepted as a viable theory.

Another effort to put quantum theory in the realist camp was put forth by Physicist Hugh Everett, who suggested superposition was not an issue of certainty and uncertainty and that all states could be real if the universe *split* at each superposition. In this way both states are real, though in separate universes with different realities. This notion became known as the Many-Worlds Theory. While that theory might be a usable interpretation of quantum weirdness, it is also unprovable, so it stays on the fringe of viable scientific theories.

Quantum mechanics has proven to be reasonably accurate and is therefore believed to represent a certain reality of nature. But it is fraught with multiple levels of quantum weirdness. Let's examine that quantum weirdness more closely, starting with the idea of uncertainty.

Uncertainty

Science is often looked upon as the search for the truth about nature. For physicists, this largely means finding the truth about matter: how it moves and interacts, and how these things change with time. Newton's observations and mathematics described things pretty well. Once scientists began exploring below the atomic level, however, the movement and interactions that changed over time were not so easily explained with Newton's laws. The belief that particles could be both point-like objects and waves, and exist in two distinct states, created a duality difficult to explain, making the quantum world appear quite different from the macro-world we live in.

Quantum Theory deals in the realm of subatomic particles, which cannot be seen, but are inferred from experimental evidence. Because we can't see the particles, we create sensitive equipment to detect them, but even that equipment has its limits. Trying to accurately measure the characteristics of tiny particles we can't see and can barely detect is a challenge. Comparatively, we are still quite distant from such things as quarks and gluons. Imagine you are a scientist trying to conduct experiments by observing people on earth from as far away as the International Space Station (ISS). Through a special apparatus one could gain an idea of what the subjects on earth were doing, but given the distance and sizes involved, one could never be certain of knowing their precise movements or the causes affecting their movements.

Bouncing something as light as a photon off a particle to gain some information about the particle can change the particle's movement or characteristics. Even accounting for the bouncing effects, we can only obtain limited knowledge from a single encounter with the particle; if we determine its exact location, we cannot know its exact momentum and vice versa.

Looking deeper into the makeup of matter, it was inevitable that we would reach a limit to what we could observe or measure given the size and speed of subatomic particles. This leaves us *uncertain* when trying to obtain two characteristics of a particle simultaneously. That limitation is encapsulated in the *Heisenberg Uncertainty Principle* (HUP), which is based on the underlying notion that objective observations of nature's most basic entities are impossible.

The uncertainty principle, developed by physicist Werner Heisenberg in the 1920s, stimulated the development of Quantum Mechanics, the mathematical formulation that gives probable answers when dealing with particles and their interactions. Quantum mechanics has since broadened into Quantum Theory that encompasses all aspects of matter and energy at the quantum level. Being so important to the foundation of quantum theory, we should take a closer look at this uncertainty principle.

The Heisenberg Uncertainty Principle (HUP)

In Quantum Theory, the HUP is a fundamental principle that explains why it is impossible to simultaneously measure more than one quantum variable, such as position and momentum or energy content and time. This principle was formulated when Heisenberg was trying to build an intuitive model of quantum physics. He discovered that there were certain fundamental factors that limited our actions in knowing certain quantities. He created a core equation that accurately expresses this limitation; the more we know about one variable, the less we can know about the other.

From QT's standpoint, for a particle to remain a 'particle' over time [*as opposed to a wave*], it must possess definite values of position and velocity, but the Heisenberg Uncertainty Principle denies this possibility, giving rise to a delocalized wave-like aspect of quantum particles.[4] In other words, since a particle cannot possess both a specific position and momentum, it possesses a wide range of both values at the same time.[5] These theoretical ideas are what makes the quantum world appear weird.

One interpretation of the HUP is to recognize it as a purely mathematical deduction. Taking measurements is a human endeavor. If when doing so we cause a particle's momentum to change, it should not be interpreted as a reflection on the particle itself. If we avoid disturbing the target particle and find we cannot then determine its precise momentum, we should not interpret *that* as a reflection on the particle either.

In other words, our inability to measure does not justify assigning to, or taking away from, variable characteristics of particles.

The HUP is an important and necessary principle that needs to be taken into consideration when predicting experimental outcomes, but the principle is associated with a human endeavor and should not be applied to the description or behavior of particles independent of that endeavor. As Physicist Lee Smolin tells us, "...the terms by which science describes reality cannot involve in any essential way what we choose to measure or not measure."[6] From this line of thinking the appropriate interpretation of the HUP is that it is a very helpful mathematical deduction that simply expresses a human limitation in simultaneously measuring two self-created particle attributes.

There are, of course, opposite opinions. Some physicists have adopted the HUP as representative of a fundamental aspect of particles themselves. If the position of a moving object is known but not its velocity, in the next instant of time, having a range of velocities available to it, the object would effectively come to occupy a range of different positions simultaneously. In other words, it would become a wave of superimposed position states.[7]

This interpretation says that objects subject to the HUP no longer occupy a specific position when their velocities are known and have no specific velocity when their position is known. When we know a particle's position, then it behaves like a particle. But it does so only instantaneously. As soon as we look away, the HUP dictates that the particle must start behaving like a wave again because its velocity is not well determined.[8] The idea of multiple states is called *superposition*. It should be noted that this interpretation requires a belief in wave-particle duality (*discussed in Chapter Three*), without which there would be no basis for the idea of superposition.

It seems more reasonable that the HUP simply expresses a human limitation and says nothing about the attributes of particles themselves. The limitation inherent in the HUP makes sense when one realizes two things. Firstly, one of the two quantum variables under consideration is in constant change. Momentum is defined as velocity times mass. Velocity means the particle under consideration is moving, i.e., its position is constantly changing.

Secondly, one must realize that there is no such thing as "now". Now is a term we use generally to refer to that point in time between past and future; it falls just after the past and just before the future. The nature of time precludes us from ever experiencing or measuring a specific "now" because before one can recognize and record it, it has already passed. Consequently, if a particle is constantly changing position due to it experiencing momentum, and there is no way to record "now" during that movement, there is no way to know with absolute precision the particle's momentum *and* position simultaneously.

One could pick a future time and say that if conditions don't change, meaning the rate of change remains constant, then one could *predict* a measurement value. This is in essence what quantum mechanics does; it predicts a value given a calculable rate of change.

Understanding Uncertainty

A thought experiment might provide some clarity as to why precise simultaneous measurements are seemingly impossible. A fan with a movie camera at a baseball game films a homerun hit over the center field fence and later she wishes to measure the ball's velocity with respect to its location. She knows the directional aspect of its velocity, pinpointed by a straight line between home plate and where the ball crossed the center field fence. From a single frame of the movie film, she can produce a still photograph that pinpoints the ball's location, but because the ball is not moving in the photograph, she cannot determine the ball's speed.

Measuring the distance the ball travels between several movie frames and knowing the timing of the frames, she can calculate the ball's average speed between those frames. Unfortunately, she cannot know exactly where the ball was in those frames when that speed was achieved. She can narrow down the number of frames, making the distance between them less and less. Even measuring the distance and time between just two frames, she will only know the average speed and not know where in that distance, no matter how small, the ball was when it achieved that speed.

She can get very close to knowing the ball's speed at a specific point, limited by the filming capacity of her camera, but no matter how good a camera she uses she will never be able to calculate both the speed and location *precisely.* On the scale of a particle's movement, the methods of calculating its location and momentum may be quite different, but I believe this gross description of the process gives the reader an idea of what physicists are up against trying to measure two aspects simultaneously.

Let's explore a little deeper why we may have such limitations and uncertainties requiring math based on probability rather than certainty. If the only method we possess to determine a particle's location and momentum is to bounce other particles off it, which may alter the target particle's characteristics, then understandably we must resort to probabilities rather than certainty. We may someday develop a different method to ascertain this information, or maybe we will find that learning both simultaneously is actually impossible, but not of significant consequence.

We may never be able to truly know or understand the minute environment in which these tiny, unseen particles reside. We know the environment is crowded and contains many different conditions, and there might even be more layers of conditions we have not yet detected. Let's take a closer look at the environment in the science lab where a test subject viewed from the ISS is under investigation to determine its location and momentum simultaneously.

The Minute Environment of a Subatomic Particle

Scientists once believed the 'vacuum' of space was nearly empty and was very much like a vacuum, but now they believe it is teeming with activity. Physicist Harold Fritzsch tells us that up close the vacuum looks more like a bubbling cauldron than empty space.[9] Physicist Kerson Huang tells us the vacuum's fluctuating electromagnetic field buffets an electron as if in a stormy sea.[10] If the vacuum is that raging, then the environment around the science lab must be a *really* turbulent place, be it in either Chicago or the International Space Station.

First, the medium in the lab is bombarded by a very high density of particles. One-hundred trillion neutrinos pass through our bodies every second.[11] Photons outnumber matter particles a billion to one, so huge quantities of near ubiquitous photons are likewise bombarding us. The medium might also contain a high density of dark matter, the most dominant source of matter in the universe. Negotiating through this blizzard of particles would be like negotiating a New York City sidewalk at 5:00 pm on a Friday. One would be challenged to maintain a straight path and constant speed with so much getting in the way.

Typical of New York City, it is bathed in a rainstorm, so the umbrellas are out. The umbrella represents a particle's field. It takes up space around all test subjects, so it is something else to either avoid or respond to as the sidewalk is negotiated, causing one to flit about relative to one's own umbrella (*field*). The fields come from all directions, and some (*electric fields*) are rather rude in their pushing and pulling, giving one at times a staggering gait. Occasionally one must absorb or emit energy packets (*photons*), altering one's momentum and challenging one's further effort to stay the course.

Then the weather turns severe. Cosmic rays (*high energy space-borne particles*) and very high energy gamma rays (*photons*) are constantly raining down at high speed. Gravitational waves of various degrees are constantly rolling through. Penergy fluctuations cause virtual particles to snap in and out of existence all around. All of this distorts the environment, making it much more like a *bubbling cauldron*.

Negotiating a straight path with constant speed under these conditions is seemingly impossible. It is no wonder a scientist on the ISS sees his test subject moving in a straight line with constant speed, but from the perspective on the sidewalk that is hardly the case. Constantly reacting to the onslaught of particles, fields, waves, forces, and energy fluctuations, one must flit about so quickly within one's field it seems like one must sometimes be in two places at once. The best satellite scientists can do is wonder why a test subject seems to be in two states, and why the scientist can only give a probability of where it will be and how fast it will be traveling at any future time.

The above analogy is not exactly a description of the subatomic environment that exists on earth, but it does demonstrate how busy the Penergy medium is and how influential and important fields are. Some things we should take away from this scenario are: (1) Particles have a field that surrounds the particle like a flexible bubble; (2) Particles move around within that bubble as demanded by the conditions in which they encounter other particles, fields, waves, and energy conditions; (3) We have no way of sensing all that is going on within the particle's minute environment; and, (4) We are left with only the ability to calculate a probability of an outcome when trying to determine such things as a particle's precise location, its scattering direction, its likely decay process, its exact energy, its tunneling capabilities, and other attributes of particles and their behavior. The minute world of the particle is difficult to measure and seemingly weird, but perhaps understandably so.

The Double-Slit Experiment

The double-slit experiment has been at the core of Quantum Theory's belief that the subatomic world being both weird and uncertain. As we discussed in Chapter Three, the double slit experiment can be reasonably interpreted as particles being both point-like and waves, establishing a particle-wave duality. But we also saw that the experiment can be interpreted as the particle's field going through both slits and causing the interference pattern, making particles solely point-like objects, just as they are always found.

Particle Entanglement

Another demonstration of particle peculiarity necessitating Quantum Theory is the topic of *entanglement*. If two alike particles originate from the same source or otherwise meet, they are said to be entangled, meaning they somehow have the means to communicate over significant distances without anything passing between them. A classic experiment demonstrating entanglement involves photon polarization. Polarization is the term used to describe the direction a photon is waving as it travels. We can't tell what direction the wave is pointing until passing the photon through a polarizer set at a certain direction. If it passes through the polarizer, we know that its wave was pointing in the same direction at which the polarizer was set.

In the basic experiment, if two photons originate from the same source and are then sent out in opposite directions toward two polarizers, the photons will do the same thing in terms of passing through the equally set polarizers. If one photon goes through a polarizer, the other photon will do the same. If the first photon is blocked at the polarizer, the other photon will also be blocked.[12] How does the second photon know at what angle the first photon was traveling in order to mimic its behavior? The presumption is that they somehow communicated that information between them just before the second polarizer was encountered.

Similar experiments have been carried out using other particle traits such as the spin-up/down characteristics of electrons, and in all cases it appears the electrons were somehow communicating with each other to statistically arrive at the correlation predicted by quantum mechanics. These experimental outcomes bothered many physicists including Einstein, who didn't like the idea of one particle affecting another without anything passing between them. He called it *spooky action at a distance*.

Entanglement allowing communication between distant particles without anything passing between them is based on some assumptions. It assumes the particles do not have any definite properties prior to being observed or measured;[13] that the properties of a particle aren't fixed until we measure them;[14] and that the polarization angle cannot be preset by the photons before starting out.[15] These are big assumptions. They are based largely on the non-realist approach consistent with the previously mentioned Copenhagen Interpretation.

Einstein believed particles could have such measurable properties, which became known as *hidden variables*. A mathematical theorem created by physicist John Bell in 1964 based on statistics and the Uncertainty Principle is said to prove hidden variables cannot be experimentally verifiable. All entanglement experiments to date have been consistent with Bell's theorem, supporting the idea hidden variables cannot be verified and likely do not exist. Experiments have never found hidden variables, but that doesn't mean they don't exist, just that they have not yet been experimentally found as defined.

Whether unobserved properties known as hidden variables exist as Einstein believed notwithstanding Bell's Theorem, may be in doubt but it has been neither proven nor disproven so their existence remains unanswered.[16] In the Tor Model, some of those hidden variables may well exist residing in the particle's field, as discussed in Chapter Three.

The idea of entanglement is being used in the development of cryptography and quantum computing and many experiments point to it being very real. Yet, perhaps there is more for us to understand. Undoubtedly scientists are finding two entangled photons miles apart having the expected properties, but when and how they acquired those properties is still an open question. It could be that when the two particles touched, they synchronized the variables in their respective *fields*. Entanglement could turn out to be a good argument for both particles producing personal fields and particles communicating through those fields. More experimentation is necessary, and we must ensure we have a full understanding of particle fields and all their aspects and potentialities before drawing final conclusions.

It is possible that our thin Penergy medium is a very dynamic form of energy capable of much more than we can possibly even imagine. It could be the means for a common inter-connectedness between all things and possibly even future cosmic communication. It is hoped that the scientific investigation into quantum entanglement will produce results suggesting how we might someday tap into that energy.

Notwithstanding all this quantum weirdness, quantum mechanics works very well. It consistently and accurately predicts experimental outcomes, though those predictions are expressed as probabilities. The original equation accounting for this ability is the Schrodinger equation containing the wavefunction. Let's examine the idea of the wavefunction and its effect on reality.

The Wavefunction

One of the issues in interpreting the meaning of quantum mechanics was a controversy regarding the meaning of one of its core equations, the Schrodinger equation. That equation contains a symbol representing a *wavefunction*. It entails the notion the particle under consideration behaves like a wave, with its measurement potentially having an infinite number of values that are evenly spread out in a wave pattern [*much like a field was described earlier*]. When a measurement is taken, the wavefunction is said to collapse producing the probability value for the attribute in question.[17] The uncertainty about the meaning of the wavefunction is expressed in a controversy over interpreting when exactly the wavefunction collapses, whether it requires a conscious observer, or even if it collapses at all.[18] That issue is known as the *measurement problem*.

The wavefunction is a mathematical term, not a real thing. No experiment has ever proven the existence of a wavefunction.[19] Quantum mechanics is accurate and is used widely throughout science and industry, consequently many see the wavefunction as something not only very meaningful, but very real. It has been elevated to the level of being part of reality, and for some it is, 'the sum total of reality'.[20] In quantum physics everything has a wavefunction; your book, the earth, even the whole universe has a wavefunction.[21] An object's wavefunction determines its behavior and the behavior of an object's wavefunction is determined in turn by the Schrodinger equation.[22] Apparently for those mathematically oriented scientists, the world and everything in it has been reduced to the status of mere elements of an equation.

Again, quantum mechanics is beautiful and accurate, but its success should not be elevated above its purpose. Being capable of giving accurate probability answers is not so mysterious. The experimental outcome determining a particle's attribute takes place when a measurement is taken and the wavefunction collapses. In the Tor Model, much of the wavefunction's success and accuracy can be ascribed to the wavefunction mimicking the particle's wavy field. To measure a particle's attribute, one must interfere with the particle, which collapses its field. When the wavefunction collapses to reveal the value of a particle's physical attribute, so does the field around the particle collapse, exposing the particle's attribute. The collapse of the wave function rhyming with the collapse of the particle's field is what makes Quantum Mechanics seemingly so statistically accurate.

We have examined quantum uncertainty, quantum superposition, wave-particle duality, quantum entanglement, and the wave-function of quantum mechanics. It is time for an assessment of whether quantum theory can be used as a basis for reality.

The Reality of Quantum Theory

Reality is the state of things as they actually exist, as opposed to an idealistic or purely mathematical idea of them. We once thought reality was just as we saw the world, but we are told that what we see is only on the surface of a deeper reality, a deeper existence. That deeper existence is described by quantum theory in terms of such things as a dual nature of particles of matter, the possibility nothing exists but fields, or the possibility we live in multiple worlds. All these notions suggest a reality much different to the one we witness daily. The enigma at the heart of quantum reality can be summed up in the expression from Physicist Sean Carroll: "What we see when we look at the world seems to be fundamentally different from what actually is."[23] Given the weirdness of these quantum ideas; can quantum theory provide us with a valid picture of reality?

There is no question quantum theory math works. The only question is whether the *interpretation* of quantum theory represents the underlying reality of how the universe is. Some of the weirdness of Quantum theory's picture of reality is self-induced. It has adopted extrapolations of the HUP, such as *physical systems do not possess attributes until those attributes are measured.* Physicist Yasunori Nomura characterizes this assertion as a metaphysical claim.[24] That means that it is not scientifically proven. It seems to be a concept left over from the Copenhagen Interpretation that was at the heart of the controversy between Niels Bohr and Einstein. Einstein of course believed that particles do have definite values for their position and momentum, and just because we can't determine them simultaneously doesn't mean they don't exist or that measuring only one allows the other to take on numerous values.

Another quantum theory extrapolation of the HUP is that if you know an electron's momentum you cannot know its location, and if you cannot know exactly where it is then it exists in several *parallel states simultaneously*.[25] The Tor Model's answer to this is that you don't know the electron's precise location because it is flitting about within its bubble-like field. In Quantum Theory the electron can be flitting about anywhere in the universe, which is an extrapolation gone too far, creating its own weirdness.

Einstein believed that quantum theory, though giving correct results, was incomplete and he did not believe a wavefunction could be used to describe reality.[26] Other physicists accept quantum theory as very much describing reality. They have extrapolated the math of the HUP and quantum mechanics into a broad spectrum of ideas including time does not exist,[27] nothing is real[28], particles don't exist but their fields do[29], very little can be known with certainty, and we probably exist in one of many worlds with similar but different histories.[30] An excellent account of all these theories can be found in Sean Carroll's, *Something Deeply Hidden*.

These ideas are principally math driven. Although math can produce accurate experimental predictions, it is a poor basis for being used to describe reality. Tim James tells us quantum mechanics is a triumph of mathematical beauty, provided you don't ask what any of it means.[31] Physicist Sean Carroll tells us that the things math proves are not true facts about the world[32], and that physicists don't know what Quantum Theory actually *is*.[33] The elements of quantum theory though accurate mathematically, may not produce a usable picture of reality for the following reasons.

- Entanglement is a very interesting phenomenon, but it may be no more than that. Experiments have proven that particles have attributes, some of which are fixed and some that are subject to variation such as the direction of spin or polarization. Experiments have also proven that two particles, when entangled by touching or having emanated from the same source, can share those attributes in statistically predictable ways. Quantum Theory says statistical predictability is due to the particles having the capacity to communicate rapidly that allows the second particle to adjust its attribute depending on the value of the first particle's attribute when measured. When and how each particle acquired the value of the attribute and whether they can communicate or preset those variables is still an open question. Particle communication without anything passing between them is a respectable theory, but so is personal fields possessing particle attributes, so the entangled particle communication theory is inconclusive and cannot yet be considered a part of our reality.

- Wave-particle duality may not be real. If future experiments demonstrate that the waviness of a particle is a trait inherent in its personal field and that it is the field that passes through the double-slit experiment causing the interference pattern, then the wave-particle duality issue may be put to rest. Wave-particle duality therefore would no longer be part of our reality.

- If particles are not waves, it means they have definite positions, though difficult to pin down within the particle's field except statistically. It also means superpositions are dubious, and that a particle seemingly in multiple locations may be explained by it flitting about within its own personal bubble-like field as explained earlier. Superposition therefore may not be part of our reality.

- Schrodinger's wave-function equation provides accurate experimental predictions and measurements. The validity of the wave-function calculations, however, could be due to the wavefunction mimicking the actual wave-like attributes of the particle's field as explained above. The wave-function therefore remains accurate and very usable, but it is simply a mathematical tool to help us understand experimental outcomes and is not a description of a deeper reality.

- The HUP is an important concept and mathematical tool but is perhaps nothing more. Without the concepts of particle duality there is no support for superposition and the HUP becomes no more than a mathematical expression of a human limitation to measurement.

Wave-particle duality, the HUP, superposition, and the wave-function predicting accurate probabilities, have all been brilliant answers to real, head-scratching problems. Those answers, however, created a picture of quantum weirdness that may have led us down mathematical roads to realities that don't exist. Imagining a reality that stems from the belief the micro-world has a dual nature and works strictly based on probability is perilous, fraught with opportunities for miss interpretation.

Lee Smolin tells us that the use of probabilities is just for our convenience and the resulting uncertainties just an expression of our ignorance.[34] Physicist Kenneth Ford points out that the use of probability in the larger world results when we don't have all the available facts.[35] That might be the case for Quantum Theory as well; we just don't yet have the means to know everything that is going on in a particle's minute environment. Physicist Nomura tells us we may never really know what is going on in the quantum world.[36]

The goal of physics is to understand the laws of nature in order to anticipate and describe the future. We strive for that goal, but we may never have a complete theory or mathematical structure to precisely describe the minute environment of subatomic particles. Consequently, we may never understand all the laws of nature that exist for them or develop a reality that includes precisely all the aspects of that micro-world. So where does that leave us in search of reality? We may have to go back to our slate of observations and deductions and see what it can tell us.

Our basic premise is that pure energy (Penergy) existed at the beginning of the universe and is the substance from which everything in the universe is made. It apparently can exist in a range of densities, which explains the creation of supermassive blackholes, the origin of the cosmic web, how particles were created, and why the universe appears so homogeneous. In a constantly thinning Penergy density may be the substance of our interstellar medium that has been expanding the universe since its origin. The Penergy medium may be a better backdrop than our current theory of undetectable, ubiquitous fields for each type of particle.

If all that is true, then particles are real, and their fields are very likely simply a disturbance in the Penergy medium. It means our reality will come from a realist point of view as Einstein insisted. If the Tor Model's interpretation of the double-slit experiment is correct, it could mean QT could be interpreted as a realist theory, and perhaps the combination of a reinterpreted Quantum Theory and General Relativity is still a viable source for reality.

A Theory of Everything

The holy grail of physics currently is finding a single theory that answers all questions. Scientists call this ultimate goal of science a Theory of Everything or TOE. String Theory is a leading hopeful, but it requires ten or eleven dimensions to exist, and there is no apparent way to prove the theory by experiment. Supersymmetry is another hopeful theory, but it requires a whole bunch of new particles that have not shown up yet in our particle colliders.

Many believe a TOE might be obtained from combining Quantum Theory and General Relativity, but combining the two has so far proven difficult if not impossible. Quantum theory is working from outside the bubble of the particle's field, which makes the accuracy of their measurements regarding the dodgy movements of the particle inside the field subject only to probability. Quantum theorists are trying to figure out a way to quantize gravity. That doesn't seem realistic absent a way to account for the effect of the spin of every individual Tor within any mass large enough to possess a measurable gravitational field, which doesn't seem likely. As Physicist Lisa Randall points out, the combination of GR and QT into a single set of equations that describes particles and their interactions might be possible, but they won't describe reality.[37]

From the perspective of the Tor Model, the pathway to a TOE is through the understanding of just two things: Penergy and spin. They are at the heart of everything in the universe. Penergy spin creates particles. Particle spin creates fields. Field spin creates attractive and repulsive forces. Forces allow particles to move, communicate, and combine, giving structure to everything that exists.

Understanding Penergy and spin, one seemingly knows how the universe works from a physical standpoint at its most basic level. We may not need to account specifically for QT, GR, or the Standard Models, as they seem to be accounted for in the underlying concepts of these core relationships.

> Prediction: Someday a talented young physicist or cosmologist will create the mathematics to support the idea of Penergy, its dynamics, and its ensuing relationships, and from that mathematics a tentative TOE will be developed.

Understanding all these relationships might give us enough substance to create a Theory of Everything, but it would not be THE Theory of Everything. That larger theory must take into account more than simply how the universe works physically; it must additionally encompass the depth and potential of Penergy itself. A complete understanding of mind, consciousness, and intelligence would be a start, but only a beginning.

Those three subjects are of course *emergent qualities* as previously defined and discussed. Recognizing emergent qualities tells us that Penergy is much more than a simple substance we call energy and the CAGI is much more than a simple concept driving evolution. Penergy and the CAGI together have brought the universe a long way in terms of evolved complexity in the form of us intelligent, technology producing, living creatures. The universe's future surely involves much more evolved complexity, with new emergent qualities we can't now even imagine. From that perspective, The Theory of Everything might even require some insight into the universe's ultimate destiny.

Does this mean that none of the monuments to human thought, exploration, ingenuity, discovery, and creativity has produced a realistic picture of reality? Have we missed seeing the forest through the trees? That might well be the case.

To develop a true picture of the universe one must start not with QT and GR, both of which are excellent tools for measuring aspects of the universe, but instead start with the dynamics of pure energy, the substance from which the entire universe is made. Penergy and the changes it has undergone is not only the basis of cosmic history, but it could also be the basis of cosmic reality as well. The dynamics of Penergy of course include the work of the CAGI, the ultimate source of cosmic change. Their work together can be encapsulated in a few lines, which I shall repeat for emphasis.

> Penergy produces particles with emergent qualities.
> Particles produce fields.
> Fields produce forces.
> Forces produce Levels of complex matter, including life.
> Life produces intelligence and technology.
> Intelligence & technology produce greater complexity.
> Greater complexity produces even greater complexity.

This is the substance and reality of the evolution of our universe thus far. There is more to come and more lines to add. The changes at each Level are becoming greater and the emergent qualities at each Level more wondrous. From this vantage point reality is much bigger than the mechanics working at either the micro or macro level of existence.

This sequence has not yet been reduced to mathematical terms, but it seems to be a pattern that lends itself to such a formulation. Such a formulation would be looking at the universe from a new and different perspective; one with perhaps a sharper focus. Looking backward in time and deeper into the levels of matter does not seem to be giving us the picture of the reality we seek. We need to broaden our horizons, perhaps exploring a shift in focus. The reductionist philosophy that has existed in physics for the past century is not really getting us where we need to be. There is support for that notion. Some physicists believe there is a quiet intellectual revolution taking place around the reductionist paradigm being false.[38]

Unpeeling the onion in the microworld won't give us a final theory, only the core of the onion, which may be nothing more than something like the core elements of the Tor Model expressed above. A final theory will come from envisioning how the outer layers of the onion are yet to grow. A change in focus from the past to the future, and from the depths of matter to the future of evolving matter may give us a better picture and a broader perspective. We will examine the prospect of a broader *perspective* in the final chapter.

Chapter Nine
New Perspectives

Our new vision of the early universe is complete. Atoms have evolved, photons have been liberated, hydrogen and helium gases have formed, stars are in the making, and the SMBHs are pulling it all together to produce the beautiful galaxies that will continue to expand into the universe we see today. It has been quite a journey, with many unanticipated twists, turns, and surprising deductions. The journey has spurred many new ideas and theories, while at the same time bringing under question some old theories.

In Chapter Eight we questioned whether all this construction has given us a vision of reality and realized that our view of reality will depend on our perspective. It's academic that things appear differently from different perspectives, but is there a certain perspective from which our view of reality is true? Or is reality only the sum of the views from all perspectives? In our further pursuit of a proper reality, let's examine some pertinent perspectives.

The Penergy Perspective

Pure energy – Penergy - has not been given its due. It is the most important substance in the universe yet is seldom mentioned and is then only referred to obliquely. It has become synonymous with the rest energy of any mass but is otherwise pretty much ignored. Scientists use the word energy often, but they are usually referring to one of the different forms of *work energy*, which is different from Penergy.

In the science community the definition of *energy* is the ability to do work. Work is the generic description for the transference of energy between various systems. Energy cannot be created or destroyed, but it can be transferred. Work energy exists in many forms, several of which are discussed below. All these forms have highly technical definitions, but for our purposes we need only a common, less technical definition, taken largely from Wikipedia. To understand Penergy and work energy we must be clear about their relationship. Let's start by examining the different types of work energy.

Radiation: Radiation energy is that which travels as particles or waves, such as electromagnetic waves. Its physical manifestation is the photon. Radio waves, infrared waves, visible light waves, and x-rays are all radiation energy.

Light: Light energy is electromagnetic radiation in the visible range. Visible light is emitted within the frequency band capable of detection by the human eye. Light energy is emitted by any object holding heat, and since all objects hold some degree of heat, all objects emit radiation, including light energy.

Heat: While temperature is the measurement of heat contained in a body, heat energy, also called **thermal energy**, is the flow or transference of energy between two bodies. It is the exchange of photon energy by atoms or molecules bumping into each other, transferring energy from a higher energy (warmer) object to a lesser energy (cooler) object.

There are three ways of transferring heat energy: conduction, convection, and radiation. *Conduction* is the transfer of heat energy through a solid, as in leaving the tip of a hot poker in a fire and allowing the heat to gradually rise up the entire length of the poker or allowing the bottom of a pan to heat up by placing it on a burner. *Convection* transfers heat energy through a gas or liquid, as in the heating of a room or the boiling of water. *Radiation* is the transfer of heat energy through space. It is that which one feels by standing in the sun or near an open fire. In all these cases, the electrons in the receiving body are gaining energy.

Chemical: Chemical energy is the energy stored in the bonds of chemical compounds, normally carried in the electrons of the atom or molecule's outer rings. That energy may be released or absorbed during chemical reactions and is otherwise released in the form of heat (*radiation*).

Nuclear: Nuclear energy is a form of potential energy in the atomic nucleus. The energy is derived from the high energy required to form the nuclei and hold it together, and that energy is released during a nuclear reaction, especially during fission or fusion.

Kinetic: Kinetic energy is the energy a body possesses by virtue of its being in motion. The amount of measurable energy depends on the body's mass and velocity. The motion can be vibrational - such as in the plucking of a guitar string; transactional - the velocity gained or lost due to the absorption/emission of a photon; or translational - the gain/loss of velocity due to a collision with another particle or mass.

Gravitational: Bodies with mass are attracted to each other by their respective gravitational fields. Gravitational energy, therefore, is the potential energy a body possesses due to its position that would allow gravity to influence its motion.

Potential: Potential energy is the energy possessed by a system due to the position of its parts. A body lifted from the earth's surface is said to have potential energy due to it being in a gravitational field and its potential to fall toward the earth's surface, whereby the potential energy would be converted to kinetic energy.

Common Traits

Let's look at these work energies a little closer to identify any common traits. *Heat* energy, whether transferred through a medium of a solid, liquid, or gas, comes down to a high-energy electron emitting a photon, which is received by a lower-energy electron, thus spreading the level of energy throughout the medium until an equilibrium is reached. Let's label this phenomenon of energy transferred through photons as **Photon Energy**. *Light* energy and *Radiation* energy are also Photon Energy, by definition. *Nuclear* energy is also Photon Energy because the energy released by the nucleus in fission, fusion, or decay may momentarily create a knot of Penergy, but it quickly resolves into particles and photons.

Gravitational energy is derived from a mass curving spacetime, creating an attractive *field* that brings the pathways around it toward the mass. Since it derives from the mass's field, let's label this phenomena **Field Energy**. *Chemical* energy is also Field Energy because it is based on the attractive/repulsive force of the electric field. That force affects the behavior of electrons in atomic/molecular interactions, which we call chemical reactions.

Potential energy possesses the *potential* for influencing the movement of a body due to its position in a field, either gravitational, magnetic, or electric. Consequently, we will call that Field Energy as well.

We've narrowed our work energies down to three categories: Photon, Field, and Kinetic. Let's examine them a bit closer.

Photon Energy

Photons are comprised of Tor particles, but due to their balanced configuration, they present no mass and are never at rest. Due to photons mimicking an electromagnetic field (*explained in Chapter Three*), they can be absorbed/emitted by charged particles. Photons are the currency used by those particles to exchange energy, providing the needed kinetic energy to a particle, atom, or molecule for it to attach to, decouple from, or otherwise interact with another particle, atom, or molecule. When a matter particle absorbs a photon, it not only changes the velocity of the particle but the calculation for the particle's total energy, as well. The value for the total energy is found in the calculation for the change in the particle's Kinetic Energy.

Photons, therefore, are a *source* of work energy because they can be absorbed, changing a particle's Kinetic Energy.

Field Energy

A particle's velocity can be changed when acted upon by a force. In the Tor Model, forces are derived from gravitational, magnetic, and electric fields, all of which originate from the spin of a mass such as a particle or blackhole. Near these fields an unincumbered particle will be *accelerated*, meaning changed in its velocity, either its speed or direction. The value of that energy is measured in the change of the particle's Kinetic Energy.

Fields are therefore a *source* of work energy because they create a force with the capacity to change a particle's Kinetic Energy.

Kinetic Energy

A body in motion is subject to Newton's First Law: a body in motion will stay in motion until acted upon by an outside force. That means that the unaccelerated motion of a body does not depend on any intrinsic physical energy; it needs no additional energy to continue moving. Leaving mass energy aside, the *energy* the body is said to possess is simply the work energy transferred to it by the force that placed it in motion. The body in motion is said to possess energy only due to its capacity to change the acceleration of another body, thus transferring some or all its *energy* by collision.

Kinetic Energy is work energy because it has the capacity to change the Kinetic Energy of another particle, but in doing so it gives up its own motion energy. Consequently, Kinetic Energy is not a *source* of energy but simply a vehicle for transferring energy that originated from either photon absorption or an encounter with a field.

Energy Summary

Physical energy – Penergy – as far as we know currently exists in only three densities: high-density blackholes; light-density Tor particles; and a very light-density interspatial medium. It is the energy Einstein referred to in his equation $E=MC^2$. Any other form of energy is work energy.

Work energy has the *capacity to do work* by virtue of its ability to change the inertia or acceleration of a mass. The source of that ability is either through the absorption of a photon or an encounter with a field. Photon movement and fields in turn derive their capacity from their relationship with the Penergy medium. Penergy's natural tendency to spin creates particles, and particles spinning within the very light-density Penergy medium create both particle movement and fields.

In summary, Penergy spin creates particles, particle spin creates fields, fields create forces that affect the movement of particles, which we measure in terms of work energy. We have mathematically found a way to measure Mass Energy (*Penergy*) and Work Energy in the same terms, but the two energies are different: Penergy is related to the content of particles, while Work Energy is related to the behavior of particles.

Penergy is ultimately the source of all physical and work energy. From an evolutionary perspective, Penergy is much more than blackholes, particles, fields and forces. It holds many hidden dynamics and potentialities that only become apparent as blackholes and particles evolve. Penergy in all its forms has always defined our universe. When it was a speck, when it began spinning into SMBHs and particles, and when it smoothed out into galaxies of stars with the intergalactic medium, Penergy has always been the size, shape, substance, and definition of our universe. An understanding of Penergy certainly brings a fresh perspective to our universe.

The CAGI Perspective

Those who theorize the ultimate destiny of the universe seem to consistently limit that destiny to three options: the universe will expand to a point and then collapse into a *big crunch*; the universe will expand to a point and then slow to a crawl, expanding gradually but never actually stopping; and finally, the universe will expand forever, until all matter is so separated that the universe becomes a cold, dark, lifeless, *abyss*. Those theories are based solely on the physics relating to the geometry of space and the close balance between the strength of the energy expanding the universe in opposition to the strength of the gravitational energy shrinking the universe. Seldom do we see a theory that encompasses the richness of our cosmic evolution and how it may play a part in our cosmic future, influencing the ultimate destiny of our universe.

There are many ways to look at evolution, but the Combination and Growth Imperative (*the CAGI*) discussed in Chapter Five is certainly a plausible theory. Its basic tenets are simple. Its premise is that the universe started with large swirls of Penergy that cascaded down to a band of single bits of left and right-spinning particles we have named Tors. Tors create fields, fields create forces, and forces create matter. Tors combined to create Tryks, that combined to create Sub-A's, that combined to create atoms, that combined to create cells, that combined to create organisms.

At each Level the existing, dominant form goes through its own evolutionary process, combining in different configurations whose survival is naturally selected by the elements within its environment. This process continues until a stable, strong, dominant form is produced that itself is capable of combining. During each Level's development, emergent qualities evolve that assist in that Level's survival, development, and ability to combine.

To learn the details of the Combination and Growth Imperative, see *Journey of the Universe – A New Perspective on its Past, Present, and Future Evolution*.[1] It explains each Level in detail and how cells combined into organisms and that at the top of the chain of organisms on earth is humankind. It then explains in detail how humankind might be the stable, dominant form of matter that ultimately combines into the Seventh Level of matter. It further explains in detail how that Seventh Level will ultimately combine into the Eighth Level. Much of this occurs while humans are living in space. It is an intriguing journey.

The reader may be having trouble imagining how some elements of humankind could possibly come together to create a larger form of matter. It is not as difficult as one might think. Looking over the history of humankind, especially the last two hundred years, the reader will understand how we are practically destined for that role and are currently very much on track to achieving it. Practically everything humankind has accomplished in its recent history has been unknowingly oriented toward humans combining into a larger entity. It is as if we are being guided by an unknown force, and we are globally doing exactly what we need to do to accomplish that combining process. The *Journey* also looks out further into the future well past the Eighth Level and speculates on what cosmic evolution might be all about. The reader will find that there may be much more in store for humankind and the universe than the typically pronounced endings of a big crunch or a slow drift to an abyss.

The Penergy Perspective has allowed us to peek inside the toolbox of the CAGI. Set against the backdrop of the Penergy medium, it appears that that the CAGI needs only three tools to carry out its work: 1) Tor particles – the substance from which to build complex matter; 2) fields – that which creates *forces* between particles that affects the interactions and relationships between them; and, 3) photons – giving movement energy to particles that provide them the means to couple/uncouple, form various phases (*i.e., gas, liquid, solid*), or otherwise transfer needed energy.

Like a fine painter meticulously applying color to her canvas, the CAGI molds matter into forms having surprising emergent qualities impossible to imagine given the apparent simplicity of the tools from which it must work. In much the way we appreciate a fine painter working with such simple tools, we are in awe of the CAGI's handiwork. An understanding of the CAGI certainly gives us a new perspective on the universe.

The Tor Model Perspective

Cosmologist Carolyn Devereux tells us that the basic scientific process is drawing rational conclusions from observations in an objective way.[2] That has been the goal of this treatise. We started at the big bang with a blank slate, but rapidly filled it with observations and deductions, taking us quickly through the earliest phases of the evolving universe. The journey required us to step off into the hazardous realm of creating new particles, which necessitated the creation of an entirely new model – the Tor Model. While developing the Tor

Model we found it to be in significant contrast to many of the tenets of the Standard Models.

The Tor Model is not meant to take the place of any other model. It is food for thought in the development of ideas and theories regarding the unknowns and mysteries of the well-developed models we are currently using. Though much of the Tor Model makes sense, it does not yet have a mathematical basis, so the ideas expressed, though possibly quite valid, may be slow to be accepted. The Tor Model nonetheless answers some of the important questions and issues plaguing both cosmology and particle physics. Here is a brief summary of some of those answers.

- **What were the initial conditions at the birth of the universe?**
  Beyond our belief the universe started from a very small speck of pure energy that for reasons unknown began to expand, there were no other initial conditions. Everything that subsequently happened, or was created, simply unfolds from changes in the density of the expanding pure energy, i.e., Penergy.

- **What is causing cosmic expansion (*The Dark Energy mystery*)?**
  Penergy in a non-spinning state naturally expands. As the early universe expanded the Penergy density thinned more and more until it is now our very thin interspatial medium, which is also the "dark energy" that has been expanding the universe since its inception.

- **How were supermassive blackholes (*SMBHs*) formed?**
  Penergy, like interstellar gas, with sufficient density naturally swirls, spins, and condenses. In the early universe when the Penergy was still very dense, clumps of it began swirling, ultimately creating a cascade of swirls. The larger

of those swirls eventually spun themselves into pure-energy SMBHs.

- **How was the Cosmic Web formed?**
  The cascade of swirls created smaller and smaller swirls at their edges, ending in very tightly wound swirls. The large swirls became blackholes and the very small, tightly wound swirls became particles, both of which had the correct spin dynamics to sustain themselves. The intermediate sized swirls, however, could not sustain themselves in the very-thin density Penergy and eventually dissipated back into the Penergy medium. Their disappearance left large voids surrounded by various densities of SMBHs that became galaxies at the periphery of the voids, creating the cosmic web.

- **Why is the universe so homogenous (*The Horizon problem*)?**
  The cascade of swirls and thinning Penergy eventually drew down to particle creation. The larger swirls were not quite developed into SMBHs, allowing the first particles of photons and matter to soar around the universe and begin combining. By the time the SMBHs were fully formed, the universe was homogeneous. The SMBH's then gravitationally pulled in the existing matter and gases, creating the billions of galaxies we see today.

- **How were the first particles created, and why are there three generations?**
  At the end of the cascade of Penergy swirls were very tiny swirls that spun into tightly wound particles. These tiny tornados are named Torons, or Tors for short. The first two generations of composite particles were unable to survive in the thinning Penergy and dissipated back into the Penergy medium. They still come into existence momentarily when there is a knot of Penergy of sufficient density, but they decay quickly into lighter, stable particles. The

third generation is the Tor composite particles that make up the quarks, electrons, neutrinos, and photons we are familiar with today.

- **What happened to the missing Anti-Matter (*The Antimatter mystery*)?**
  The anti-matter isn't actually missing. Those particles are part of an array of constituent elementary particles that make up the subatomic particles of today. Their presence is necessary for the creation of opposite charges whose attractive/repulsive properties are necessary for the creation of both fields, forces, and matter.

- **What is dark matter?**
  No one knows for sure. Perhaps it is particles left over from earlier evolutionary phases. Each of the first two generations, and possibly a few generations before them, went through its own evolutionary process of combining into different configurations until a strong, stable, combination was formed. The sea of remnants of those evolutionary trials were combinations that were strong enough to survive, but not adaptable enough to become a part of something larger. They would float around, bumping into each other until their energy and charges were neutralized, rendering them "dark matter".

- **Why the disparity in the theorized level of vacuum energy and the actual measurement; a difference in the order of 10^{120} (*The Cosmological Constant problem*)?**
  According to quantum physics, the vacuum of space is awash in virtual particles constantly popping in and out of existence and having a theorized vacuum energy density of 10^{105} Joules per cubic centimeter. In the Tor Model, virtual particles only come into existence due to a fluctuation in the medium creating a small knot of Penergy, such as near blackholes and other large masses, or within the environ-

ment of a planet's atmosphere, such as our own. The measure of that limited energy density is likely much closer to the actual energy measured, which is only 10^{-15} Joules per cubic centimeter.

- **What are emergent qualities?**
  Emergent qualities, as used in the Tor Model, are special attributes of matter that are unique to each evolutionary Level. They seem to evolve to assist in the development, survival, and combining capability of each new Level, though viewing the apparent simplicity of the Level of matter at the start of its evolutionary process, one would not expect such attributes to evolve.

The Tor-Model is rich in new ideas, many of which conflict with current theory. Perhaps some of those ideas will be the source of new theories that give both Standard Models a better foundation. As noted by Astrophysicist Stuart Clark, "When things have looked impossible before, the breakthrough has finally come when a brave scientist has thrown away a cherished assumption. Often this assumption has become so entrenched that many think of it as established fact."[3] Perhaps throwing out the cherished assumption that our complex particles were created within the first minutes of the Big Bang will be the breakthrough the Standard Models needs to get on a better footing.

If Penergy is the stuff from which everything is made, and the CAGI is the story of how that Penergy has and will change, then the Tor Model is simply a theory of how those changes may have taken place in the early universe. The model goes to the core of exposing the relationships between Penergy, fields, forces, and matter. An understanding of the Tor Model certainly gives us food for thought and a new perspective on the universe.

The Mathematical Perspective

Physicists view matter and energy and the relationships between their various manifestations as having an inherent mathematical basis. It is believed or at least hoped that someday a single equation, or series of equations, will be found that simply and succinctly describes how everything is connected – a Theory of Everything, or TOE.

Expressing a TOE in a mathematical form will not be easy because the universe is constantly changing in ways we cannot anticipate. The natural core substance of the universe, Penergy, has not revealed all its secrets. Through the CAGI, evolutionary changes are taking place universe-wide and not all at the same time. With the evolution of each Level of matter, more emergent qualities will evolve having a significant impact on the appearance, capacity, and novel attributes of that Level and its progeny. The presence of a new Level of matter affects the environment in which it resides, and the emergent qualities often have a substantial impact on determining what those effects will be.

Much of the exploration into how the universe works has been oriented toward a theory combining the four forces and peeling away matter to get to its core. This reductionist approach is fine for core particle discovery, but it will not by itself give us the picture we need for a theory of everything unless we wish to confine the scope of that picture solely to the mechanics of the universe. Exposing the forces and building blocks and their behavior and relationships is only a small part of the bigger picture. The math of General Relativity and Quantum Mechanics is brilliant and accurate, but it only gives us measurements and expected experimental outcomes, not an actual picture of the universe or how it works and evolves in the larger picture. Melding the two together isn't likely to solve that problem.

A NEW VISION OF THE EARLY UNIVERSE

According to Physicist Michio Kaku, a theory of everything will have an infinite number of solutions depending on the hand-picked initial conditions chosen.[4] Aside from the belief the universe started from a high-density speck of pure energy that began expanding, there is little need for any other initial conditions. The simpler we keep the initial conditions, the less problematic the consequent explanations, theories, and mathematics. The problem with starting with additional initial conditions is that the equations that are then applied to describe those conditions and everything that follows must be consistent with those first equations. That creates a mathematical bias that limits our field of view and leads to theoretical difficulties or dead ends.

If a single equation is to describe everything, it would seem to necessitate on one side of the equation a symbol for Penergy, the substance from which everything is made. Perhaps it should also have a symbol for the CAGI - the ultimate description of how the universe has changed and will continue to change in the future. The symbol for the CAGI might include the concepts of *entropy* and *emergent qualities*, as these are essential for understanding evolutionary change.

Mathematics is essential to our understanding, development, and existence. It is the best language for accurately describing the universe and is very important in many ways. Our endeavor to find the terms to describe the entire universe is certainly worthwhile. The sequence of the changes Penergy has gone through mentioned at the end of Chapter Seven may someday be reduced to a mathematical formulation representing at least a part of the theory of everything. An understanding of that new mathematical formulation, or an equally valid theory of everything, will certainly give us a broader perspective on the universe.

The Life Perspective

Humans naturally think of *life*, especially human life, as something special. We are conscious, intelligent beings, with the extraordinary capability of understanding the universe. The creation of life out of inanimate materials is nothing short of miraculous. We have a soul. We know God. Of course, we are special.

Perhaps we are. But perhaps we are simply complex organisms representing the earthbound Sixth Level of matter (1st-Tors, 2nd-Tryks, 3rd-Sub A's, 4th-Atoms, 5th-Cells, 6th-Organisms). We are endowed with a level of consciousness and intelligence that no previous Level seems to possess. But these are simply emergent qualities inherent in our Level of evolution. Every Level has its own emergent qualities. The Seventh Level will be created when we humans combine (*if we survive without annihilating ourselves*). That Level will have its own emergent qualities compelling them to feel superior. They will define their existence in a term more appropriate to their attributes, and so endowed believe they are "special". It sounds a bit egotistical, but that is a good thing. The belief in their superiority will keep future Levels working to better their lot, keeping them motivated to develop the resources and means to create the next Level.

The special existence of humans carries over to the *anthropic principle* – the idea that life is special in the universe. The essential elements in the universe - matter, energy, forces, and their relationships - can all be measured and reduced to about thirty numbers called universal constants. Viewing those constants, it has been noted that they appear to be fine-tuned for the existence of life and that if we change one of the constants by a relatively small amount, we find that it makes the universe inhospitable to life.[5] Why is the specifications for the universe so hospital to life? This is known as the *fine-tuning problem*. It is one of nature's deepest mysteries.

A belief that the universe was designed to suit life seems compelling. Stephen Hawking recognized the argument, and after listing all the elements that had to be just right for us to exist, he says, "Our universe and its laws appear to have a design that is both tailor-made and support us, and if we are to exist, leaves little room for alteration."[6] Physicist Jim Baggott, however, reminds us we are very much a naturally evolved part of this reality, but we are not the reason for it.[7]

In the bigger picture, we humans are merely a product of the universe due to evolutionary natural selection. It's not that the universe created the conditions specifically for 'life' to exist, it's that the CAGI simply exploited those tenuous conditions and made life from them.

There are over 100 billion galaxies, each with 100-200 billion stars, each with an untold number of planets and moons. If even a small fraction of them is suitable for the biological experiment of life, it means the odds are we are not alone in the universe. If the CAGI theory is valid it would mean there are Level Fives, Sixes, and perhaps Sevens throughout the universe and we are definitely not alone.

Alone or not, we are the trustees of the planet earth, and it's quite possible we are also a rare form of intelligent life. As a global society we need to acquire a new image of this bigger picture of who we are and what our responsibility is. We can no longer afford to act like school-yard children squabbling over possessions, territory, rights, rules, what God said, or whose God is supreme. Given the devastating technology we hold in our hands, that behavior is just not acceptable.

As a global society, we need to adopt a new perspective on *life*, what it means, and how we are going to respect it. If we are indeed special, we must realize the responsibility that comes with that privilege. Our primary goal must be to survive. That means we must take care of our planet, ourselves, and plan for the future. Assuring our survival as a species will mean someday leaving our planet and building colonies in space and on other planets or moons. This feat is best accomplished as a global effort. Creating such a global image, uniting in the common goal of manifesting our destiny in space, and working together to create a larger, more complex organism, will certainly give us a new perspective on the concept of *life* in our universe.

The Cosmic Perspective

It is apparent that a meaningful reality is more than simply understanding the mechanics of how matter and energy work on a micro or macro level. It is more than understanding how the universe has evolved, or how life has evolved. It is also apparent that reality depends on one's perspective and to understand the true reality of the universe – the cosmic reality - we must encompass many perspectives. Because the universe is forever growing in complexity in unanticipated ways due to the CAGI, a complete picture of a cosmic reality may not be known until we are much closer to the manifestation of the universe's purpose.

It is also apparent that the destiny of the universe is far from it ending in either a big crunch or a cold abyss. While much of our cosmic history has been structured by the persistence of the CAGI and refined by natural selection, there are far too many wondrous things evolving for this complex cosmic dance to have been choreographed by mere chance. The creation of emergent qualities alone tells us there is much more to the capacity of the Penergy and the creativity of the CAGI than we can even imagine. The universe is surely on a journey. The most incredulous thing about it is that humankind may be in the envious position of being capable of figuring out its mysteries, as well as playing a small role in its destiny.

The persistence of the CAGI and its capacity for developing spectacular emergent qualities tells us there are fantastic experiences in our future. We need only to survive to become part of that cosmic destiny. To guide us in such an endeavor we need a picture of, or at least a vague image of, a cosmic reality. What picture of cosmic reality can we derive from the lessons of the many perspectives we have examined so far?

A Cosmic Reality

The many perspectives we have examined tell us there is a much broader vision of the universe than we have glimpsed thus far. The Tor Model has given us a vision of how the universe may have started and evolved, but more importantly it has given us an introduction to Penergy and the CAGI. These two things are essential to understanding the universe.

Penergy is more than simply pure energy that creates and moves matter; it is the dynamic substance from which is molded the many unique qualities of our entire universe. Cosmologist Lawrence Krauss in summing dark energy says, "It is natural to suspect that its nature is tied in some basic way to the origin of the universe. And all signs suggest that it will determine the future of the universe as well."[8] Dr. Krauss's suspicions are right on. Dark Energy (*Penergy*) has everything to do with the universe's unfolding, evolution, and destiny. Penergy is magical, and for us the prospect of witnessing more of its exciting creations and surprises is titillating.

The CAGI is the force that molds Penergy's unique qualities, bringing into existence fields, forces, relationships, complexity, intelligence, and life. It will mold future combinations of matter into forms we can now only imagine. It will create complex emergent qualities of which we shall be in awe.

Life is the Level of matter possessing the intelligence and technology to mold matter at an accelerated rate, transcending natural selection in its evolutionary process. Natural selection will remain a factor, but from the Sixth Level on, the Level of matter itself will be partially responsible for engineering its form, features, and future survivability. The definition of "life" will surely be redefined as new Levels of matter are created.

In addition to being the descriptive language of the universe, mathematics will be the array of tools we will forever need for material development and problem solving to secure our survival.

From these perspectives it is obvious that we are on a wonderous journey. We are a life form possessing the intelligence and technology to influence our evolutionary development, allowing us to move off our planet to secure our future. In hindsight, our technological development has been unknowingly pushing us in that direction for the past two hundred years. That technology is allowing us to modify ourselves in preparation for adapting to space and develop the physical means to live there.

We may never know the exact history of how the early universe evolved, but while examining that realm we have learned a great deal about cosmic evolution and our own possible future. To realize that future we may have to shift the definition of reality away from the mechanical workings of the universe toward the bigger picture of the universe's changing evolution.

In the backdrop of Penergy and the CAGI, we are products of an evolutionary journey that is inexorably building something very special. That something may very well be comprised of intelligence and technology, both of which we possess, creating the prospect of our playing a small part in that ultimate destiny. We know not the specifics of our immediate future, but our limited perspective suggests the overall course we must undertake to be a part of the cosmic destiny is for us to survive, to explore, and to evolve. The lessons we've learned from our different perspectives provide the knowledge, blueprint, direction, and means to accomplish that goal. We are not only sightseers on this wondrous journey; we are a part of the journey itself.

*   *   *

Thank you for joining me on this wondrous journey. It has been a wild ride and one that leaves us with a vision of the universe from a unique perspective. A perspective that only comes about if we are daring enough to view the universe not only from its mathematical description, but from an evolutionary basis derived from our observations and deductions. It is hoped that this new vision provides a better foundation not only for understanding the journey of the universe, but the journey of humankind, as well.

It is a vision much different from today's accepted vision of the early universe, but I have faith that some of the truth espoused herein will eventually seep into the scientific consciousness. As Physicist Luciano Rezzolla tells us with regard to an important lesson that history has taught us, "What appears exotic and obtuse today can become plausible and acceptable tomorrow."[9]

Physicist Don Lincoln in his new book, *Einstein's Dream*, points out that to find a TOE we will have to do it the old fashion way – by studying the world around us. He says, "Occasionally we will notice a loose thread in our theories and find that a tug on it unravels the entire fabric of our understanding of the laws of nature, and we will then use that loose thread to reweave it into a newer and more beautiful tapestry, one that better represents the way the universe actually works."[10] Perhaps the Tor Model espoused herein is that thread.

NEW PERSPECTIVES

Glossary

Accretion: The accumulation of particles into a massive object by gravitationally attracting matter, typically gaseous matter, in an accretion disk. Most astronomical objects, such as galaxies, stars, and planets, are formed by accretion processes.

Alpha Particle: A composite of two protons and two neutrons simultaneously ejected from a nucleus during nuclear decay.

Annihilate: A conversion of mass to pure energy when a particle of matter meets its anti-matter particle. The amount of energy involved is governed by Einstein's equation $E=MC^2$.

Anthropic Principle: The idea we can make theories about the universe and the laws of physics based on the fact we exist in the universe.

Anti-matter: The twin of a matter particle with the same mass, but opposite spin/charge. When particles are created in pairs, one is matter and the other anti-matter, so all matter particles have an anti-matter twin. The two annihilate when they meet.

Baryon: Particles constructed of three quarks. Protons and Neutrons are baryons.

Beta Decay: The decay of a down quark into an up quark, electron, and neutrino.

Big Bang: The Standard Model's version of the origin of the universe from a small concentration of energy out of which space, time, and matter were abruptly created.

Blackhole: An object with a gravitational field so dense that not even light can escape it. Anything falling inside the black hole's outer boundary, called the *event horizon*, cannot escape. See also, *pure energy black hole*.

CAGI: The Combination and Growth Imperative that is driving the evolution of the universe.

Charge: See Electric charge.

Cosmic Microwave Background: The CMB is a relic of atom creation when photons were liberated. They are seen today as a low-level electromagnetic field pervading all space and measured to be 2.7 degrees Kelvin rather consistently.

Cosmology: The study of the origin, structure, and evolution of the universe.

Dark Energy: The energy theorized to be driving the expansion of the universe. There is no evidence for it other than it explains the

GLOSSARY

expansion, but its source and nature are unexplained. The author believes dark energy is the thin-density, pure energy left over from the Big Bang after early blackhole and particle production consumed most of the denser pure energy.

Dark Matter: Unseen matter having no electromagnetic signature and therefore detectable only due to its gravitational influence. What it is made of is unknown. The author believes it may be made up of particle fragments that were left behind in the particle creation era after the Big Bang.

Density: The ratio of a particular quantity, such as energy, to the volume in which it is contained.

Diffraction: The process by which waves are spread out as a result of passing through a narrow aperture or across an edge, typically accommodated by interference between the wave forms produced.

Dynamic Gravity: The author believes that gravity is generated directly by the spin of a black hole affecting the area outside and just inside the blackhole. Dynamic gravity affects the area deep inside the blackhole. It does not take effect until the black hole has matured, processed matter back to pure energy, and processed that pure energy back into a very dense pure energy. It is the very dense pure energy at its core that drives up the spin rate of the black hole that, in turn, shrinks the event horizon.

Electric Charge: An attribute of electrons and quarks that gives them attractive and repulsive qualities. According to the Standard Model, the electric charge force is created by the respective particles exchanging virtual photons. According to the author's model, the electric charge force is created by a differential in the Tor particle count between the respective particles.

Electric Field: The field created by a charged particle or particles.

Electromagnetic field: The interconnected electric and magnetic fields produced by an electric charge or photon in motion.

Emergent Qualities: Properties and patterns not explainable by reducing a complex system to its simplest parts; behaviors that cannot be anticipated by viewing a system's sub-parts in isolation.

Energy: The ability to do work. For a full explanation, see Chapter Nine, *The Penergy Perspective*.

Entanglement: A quantum theoretical phenomena wherein a pair of particles interact and then once divided seem to have the capacity to communicate, even at large distances, without anything passing between them.

eV: *see MeV*.

Event Horizon: The edge of a black hole that serves as the point at which if anything passes, it cannot escape the black hole's intense gravity.

Fermion: A generic term for particles having ½ spin, which includes quarks and leptons.

Field: An area of Penergy medium that is directly influenced by the vibrational presence of a particle or group of particles.

Galaxy: A loose gathering of stars and interstellar gases and particles held together by gravity.

Gamma Ray: A very high energy photon.

General Relativity: Einstein's theory of gravity, which cannot be distinguished from acceleration, causes a curvature of space-time, equating it with the geometry of space.

Gluon: A virtual particle that according to the Standard Model mediates the strong force holding quarks together.

Gravitational Field: According to General Relativity, the field created by the presence of mass. In the author's model, a gravitational field is created, like all other fields, by particle or blackhole spin and its effect on the surrounding Penergy medium.

Gravitational Wave: Oscillations in space-time caused by the movement of mass. Generally, they are undetected unless the movement is substantial, such as the collision of heavy objects such as black holes or neutron stars.

Graviton: The Standard Model's theoretical particle believed to be the messenger particle of a quantized gravitational field.

Hadron: The generic name for particles made up of quarks. Protons and Neutrons are Hadrons.

Heisenberg Uncertainty Principle: The idea that the more accurately we measure a particle's position, the less we know of its momentum, and vice versa. A similar relationship holds for energy and time, as well as various other observables.

Homogeneity Problem: Inability of the Standard Model's big bang theory (*without inflation*) to explain the present-day homogeneity of the universe. In the big bang theory, particles in the universe would not have had an opportunity to interact and reach thermal equilibrium.

Infinite: The notion something can go on forever, without limit or boundary.

Inflation: A theory of exponential expansion of the universe occurring within the first second after the big bang. It was developed in order to overcome the *homogeneity* problem and other cosmological issues.

GLOSSARY

Initial Mass: a measurement of the resistance to the acceleration of a body responding to a force.

Interference: Combination of two waves with the same frequency leading to reinforcement or diminution of joint intensity depending on the phase of the waves when meeting. Waves in phase meet constructively, and waves 180 degrees out of phase, destructively.

Interspatial Medium: The substance pervading the universe making up the space between stars and galaxies.

Kinetic Energy: Energy said to be possessed by a body in motion.

Large Hadron Collider (LHC) is a particle collider located near Geneva, Switzerland, and is the largest of its kind, being a ring twenty-seven kilometers in circumference. It smashes particles together at high-speed allowing scientists to theorize from the collision debris how particles are constructed, relate, and behave.

Light Year: The distances traveled by light in one year: 9.46×10^{12} kilometers.

Magnetic Field: The field created by the alignment of charged particles, that in turn causes other charged particles to align.

Meson: A composite particle made of a quark and an anti-quark.

MeV: Million electron Volts. A measurement used often by physicists to describe the mass or energy of a particle. An electron volt is the amount of energy acquired by one electron passing through an electric field having a one-volt differential. A MeV is 10^6 eV and A GeV is 10^9 eV.

Mirror-image: the view of an object as if it were a reflection in a mirror. That reflection would be opposite that of the object, but otherwise be the same in every visual respect.

Molecule: A stable combination of two or more atoms held together electromagnetically by the sharing of one or more electrons. A molecule has properties different from any of its atomic constituents.

Mutation: A mutation is an alteration in the nucleotide sequence of DNA, often caused by exposure to ultraviolet radiation.

Natural Selection: The process whereby organisms better adapted to their environment tend to survive and produce more offspring, perpetuating those survival traits. In the case of elementary particles, the best composite combinations tend to survive the frenzied environment and go on to become dominate and available to combine again.

Neutron Star: Late stage in the life of a star, reached when portions of the outer layers collapse under the influence of gravity. They

GLOSSARY

form a dense inner core of neutrons created by protons and electrons forced together under extreme gravitational pressure.

Nucleus: The small, positively charged center of an atom comprised of protons and neutrons.

Pauli Principle. That principle encompasses the fact that no two like fermions (*electrons or quarks*) can occupy the same state, meaning carry on the same function. It is the reason that two electrons cannot occupy the same state in an atom. However, two electrons can appear in the same orbital ring if one is oriented with its spin pointing up, while the other is oriented with its spin pointing down.

Penergy: According to the author's model, Penergy is short for *pure energy* and is another name for raw energy, which is the energy from which everything in the universe is made.

Pion: a composite particle made up of two quarks believed to be a part of the force holding the nucleus of atoms together.

Polarization: The direction the amplitude of a wave is displaced.

Positron: The positively charged anti-particle to the electron.

Pure Energy Blackhole: The author's notion that the super-massive black holes at the heart of galaxies were created very early and spun out of pure energy.

Quantum Field Theory (QFT): A theoretical framework that combines classical field theory, special relativity, and quantum mechanics. QFT treats particles as merely manifestations of their fields.

Quantum Mechanics: A mathematical formalism that gives *probability* when calculating a particle's aspects.

Renormalization: a method used in quantum mechanics to remove unwanted infinities from equation solutions.

Singularity: A point in space-time when a physical quantity, such as mass or energy, becomes infinite.

Special Relativity: Einstein's theory relating space and time based on, 1) the laws of physics are invariant in all inertial frames of reference (*reference frames with no acceleration*), and 2) the speed of light in a vacuum is the same for all observers regardless of the motion of the light source or observer.

Spin: In general, and in the author's Tor Model, spin is the quality of something turning on its own axis. In particle physics it is a fundamental property of elementary particles that describes their state of rotation in terms of angular momentum times h/c^2. The h is Plank's constant: 6.626×10^{-34}, and c is the speed of light: 3.0×10^8 m/s.

GLOSSARY

Standard Model: Here used to represent the combination of the Standard Model of Particle Physics and the Cosmological Standard Model (*also known as the Concordance Model*) that together spell out the origin and history of particles and the universe.

Sub-A's: This is short for Subatomic Particles, referring to those Standard Model particles smaller than an atom.

Supermassive Blackhole (SMBH): Blackholes that are thousands, millions, and billions of times bigger than our sun. These huge blackholes are at the heart of every galaxy.

Supernova: The explosion of a dying star, spewing its heavy atoms out into space.

Supersymmetry: A theory of particle creation wherein all particles known today have a cousin particle much alike in appearance providing eloquent answers to particle issues, however, the cousin particles have never been seen and have not shown up in LHC experiments.

Symmetry: The ability of a system to appear unchanged despite undergoing some transformation.

Thermal Energy: is the flow or transference of energy between two bodies.

Thermal equilibrium: A state of uniform energy, or heat, throughout a given space, surface, or body.

Tor Model: The author's model of particle creation and the evolution of the early universe.

Tor: According to the author's Tor Model, a Tor is the first level of matter; the most elementary particle spun directly out of pure energy, and the particle from which all others are made.

Tryk: According to the author's Tor Model, a Tryk is the second level of matter, a combination of Tors and Tor-Chains, and the particles from which quarks and leptons are made.

Virtual Particle: A short lived particle created from a knot of energy occurring in the interspatial medium through natural fluctuations in the energy field. The knot of energy immediately begins to deteriorate, so it is insufficient to retain the particle in the current energy density.

Wave-Particle Duality: The theory that particles are both waves and point particles, displaying characteristics of both depending on how they are perceived or measured.

Notes and References

Introduction

1. Devereux, Carolyn, *Cosmological Clues – Evidence for the Big Bang, Dark Matter, and Dark Energy*, 2021, p.72.
2. Devereux, Carolyn, *Cosmological Clues – Evidence for the Big Bang, Dark Matter, and Dark Energy*, 2021, p.3. A further explanation for the observational evidence supporting the Big Bang Theory can be found in Chris Impey's *How it Began*, p.274-276.
3. Mersini-Houghton, Laura, *Before the Big Bang – The Origin of the Universe and What Lies Beyond*, 2022, p.6.
4. Impey, Chris, *How It Began – A Time Traveler's Guide to the Universe*, 2013, p.298.
5. Cliff, Harry, *How to Make an Apple Pie from Scratch*, 2021, p.302.
6. Stuart Clark, *The Unknown Universe – A New Exploration of Time, Space, and Modern Cosmology*, 2016, p.287.
7. Lindley, David, *The Dream Universe -- How Fundamental Physics Lost its Way*, 2020, p.94.
8. Randall, Lisa, *Knocking on Heaven's Door*, 2012, p.412
9. Lindley, David, *The Dream Universe -- How Fundamental Physics Lost its Way*, 2020, p.56.

Chapter One
The Origin of SMBHs & The Cosmic Web

Initial Observations

1. Devereux, Carolyn, *Cosmological Clues – Evidence for the Big Bang, Dark Matter, and Dark Energy*, 2021, p.73. Dr. Devereux says that the universe started from a single point, a singularity, and that all the energy that exists today, existed at that point.
2. Jones, Mark, Robert J.A. Lambourne, and Stephen Serjeant, Editors, *An Introduction to Galaxies and Cosmology,* Second Edition, 2015, P.210. The authors say, "In the simplest cosmological models that are consistent with the cosmological principle it is usually imagined that the universe is completely filled with a uniform gas or fluid." That imagined universe fits perfectly with the model we are building.
3. Still, Ben, *Particle Physics Brick by Brick – Atomic and Subatomic Physics Explained in Legos,* 2018, P. 34.

NOTES & REFERENCES

4. Ball, Philip, *Beyond Weird – Why Everything you Thought you knew about Quantum Physics is Different*, 2020, p.115. Dr. Ball more precisely says that in the earliest moments of the Big Bang the entire universe was smaller than an atom, which he characterizes as a quantum-mechanical entity.
5. Tyson, Neil deGrasse and Donals Goldsmith, *Origins – Fourteen Billion Years of Cosmic Evolution*, 2014, p.25.
6. Hawking, Stephen, *A Brief History of Time*, 1988, p.50. Dr. Hawking pointed out that the math was compelling and therefore the work became generally accepted and that nearly everyone assumed the universe started with a Big Bang singularity. He went on to say that it was ironic that he changed his mind and was trying to convince other physicists that there was in fact no singularity at the beginning of the universe.
7. Devereux, Carolyn, *Cosmological Clues – Evidence for the Big Bang, Dark Matter, and Dark Energy*, 2021, p.73.
8. Impey, Chris, *Einstein's Monsters – The Life and Times of Black Holes*, 2019, p.19.
9. Kaku, Michio, *The God Equation - The Quest for a Theory of Everything*, 2021, p.57.
10. Impey, Chris, *Einstein's Monsters – The Life and Times of Black Holes*, 2019, p.14.
11. Kaku, Michio, *The God Equation, The Quest for a Theory of Everything*, 2021, p.117.
12. Devereux, Carolyn, *Cosmological Clues*, 2021, p.3. Dr. Devereux says that the Big Bang contained all the energy of the universe, and that within the first millionth of a second, all of the basic elements and forces that we now see in the universe were formed producing our subatomic particles.
13. Seife, Charles, *Alpha & Omega – The Search for the Beginning and End of the Universe*, 2003, P.216.

The Creation of Matter

14. Gribbin, John, *In the Beginning – After COBE and before the Big Bang*, 1993, p.219.
15. Tucker, Wallace H., *Chandra's Cosmos*, 2017, p.57. Dr. Tucker tells us that Dark Energy has the peculiar property of having negative energy. He points out that normally when a gas expands, its energy decreases. Dark energy does just the opposite: the amount of dark energy keeps increasing as the universe expands.
16. Randall, Lisa, *Dark Matter and the Dinosaurs – The Astounding Interconnectedness of the Universe*, 2015, p.39. Speaking of the Hubble Constant, Dr. Randall says, 'It is a constant in the sense that today, its value in space is everywhere the same. But

actually, the Hubble parameter is not constant. It changes with time. Earlier in the universe, when things were denser and gravitational effects were stronger, the universe expanded far more rapidly than it does today.' I infer from her statement that 'things were denser' means the energy density was greater, and that made the universe expand faster than it does today.

Blackholes

17. Impey, Chris, *Einstein's Monsters – The Life and Times of Black Holes*, 2019, p.150.
18. Hawking, Stephen, *Blackholes and Baby Universes*, p.75.
19. Rezzolla, Luciano, *The Irresistible Attraction to Gravity – A Journey to Discover Black Holes*, 2023, p.107.

Supermassive Blackholes

20. Impey, Chris, *Einstein's Monsters – The Life and Times of Black Holes*, 2019, p.146. Dr. Impey says that observational evidence suggests that SMBHs a billion times the mass of our sun formed during the first billion years after the BB. "That seems to be at odds with a slow, methodical progression from small to large galaxies by mergers. It's also at odds with the maximum rate at blackhole can grow as defined by Arthur Eddington a century ago." The authors of *An Introduction to Galaxies and Cosmology* (pages 154-155) tell us that if a stellar collapsed blackhole was fed constantly at the maximum rate it would take about 750 million years to reach SMBH size. Again, not likely the process that created a SMBH in every galaxy.
21. Impey, Chris, *Einstein's Monsters – The Life and Times of Black Holes*, 2019, p.103.
22. Impey, Chris, *Einstein's Monsters – The Life and Times of Black Holes*, 2019, p.146.
23. Impey, Chris, *How it Began – A Time Traveler's Guide to the Universe*, 2012, p.165. Dr. Impey says that computer simulations show that the merger process scrambles the near circular orbits of the stars in the two [spiral] disks and scatters them into a near spherical cloud.
24. Impey, Chris, *How it Began – A Time Traveler's Guide to the Universe*, 2012, p.152.
25. Jones, Mark, Robert J.A. Lambourne, and Stephen Serjeant, Editors, *An Introduction to Galaxies and Cosmology*, Second Edition, 2015, P.92.

NOTES & REFERENCES

Alternative Theory for SMBH Creation
26. Barnes, Luke A. & Geraint F. Lewis, *The Cosmic Revolutionary's Handbook – Or: How to Beat the Big Bang*, 2020, p.238.
27. Impey, Chris, *How it Began – A Time Traveler's Guide to the Universe*, 2012, p.196.
28. Sutter, Paul S., *Your Place in the Universe – Understanding Our Big Messy Existence*, 2018, p.162.
29. Sutter, Paul S., *Your Place in the Universe – Understanding Our Big Messy Existence*, 2018, p.165.
30. Gribbin, John, *In the Beginning – After COBE and before the Big Bang*, 1993, p.212-213.
31. Impey, Chris, *Einstein's Monsters – The Life and Times of Black Holes*, 2019, p.143.
32. Impey, Chris, *Einstein's Monsters – The Life and Times of Black Holes*, 2019, p.233.
33. Tucker, Wallace H., *Chandra's Cosmos*, 2017, p.61.
34. Seife, Charles, *Alpha & Omega – The Search for the Beginning and End of the Universe*, 2003, P.105.

Chapter Two
The Origin of Particles & Dark Energy

Early Elementary Particles
1. Cliff, Harry, *How to Make an Apple Pie from Scratch*, New York, 2021, P.250.

The Mystery of the Missing Anti-Matter
2. Tyson, Neil deGrasse and Donals Goldsmith, *Origins – Fourteen Billion Years of Cosmic Evolution*, 2014, p.26.
3. Kaku, Michio, *Physics of the Impossible – A Scientific Exploration into the World of Phasers, Force Fields, Teleportation, and Time Travel*, 2008, p.247. In his discussion as to whether our universe could have been created from nothingness, Dr. Kaku says that many physicists have pointed out that it is astonishing that the total amount of positive charges and negative charges in the universe comes out to be exactly zero, at least to within experimental accuracy.
4. Cham, Jorge & Daniel Whiteson, *We Have No Idea*, 2018, p.212.
5. Close, Frank, *Antimatter*, 2018, p.84
6. Schumm, Bruce A., *Deep Down Things -- The Breath-Taking Beauty of Particle Physics*, 2004, p.124.
7. Kaku, Michio, *Physics of the Impossible – A Scientific Exploration into the World of Phasers, Force Fields, Teleportation, and Time Travel*, 2008, p.182. In his discussion on the prospects of build-

ing an antimatter rocket, Dr. Kaku describes the construction of a 'Penning Trap' containing liquid Hm'
elium and nitrogen that would store a trillion anti-protons. He says parenthetically, 'At very low temperatures, the wavelength of the anti-protons is several times longer than the wavelength of the particles in the container walls, so the anti-protons would mainly reflect off the walls without annihilating themselves.'

The Origin of Dark Energy

8. Devereux, Carolyn, *Cosmological Clues*, 2021, p.65. Dr. Devereux admits scientists do not know what dark energy is, or how it can even exist, but she is confident it does exist.
9. Devereux, Carolyn, *Cosmological Clues – Evidence for the Big Bang, Dark Matter, and Dark Energy*, 2021, p.73.
10. Randall, Lisa, *Knocking on Heaven's Door – How Physics and Scientific Thinking Illuminate the Universe and the Modern World,* 2012, p.99.
11. Livio, Mario, *The Accelerating Universe – Infinite Expansion, the Cosmological Constant, and the Beauty of the Cosmos*, 2000, p.126.
12. Baggott, Jim, *Origins – The Scientific Story of Creation*, 2015, p.20. Dr. Baggott relays this fact incidental to discussing an experiment (*the Gravity Probe*) developed to prove Einstein's theory of General Relativity by proving the earth is bending the spacetime around it.
13. Cham, Jorge & Daniel Whiteson, *We Have No Idea*, 2018, p.90.
14. Cham, Jorge & Daniel Whiteson, *We Have No Idea*, 2018, p.103. The authors make a good argument for what they call *space goo* as being the something that exists in the vacuum of space, and the something that gives space its physical nature and the ability to ripple, bend, wave, and expand.
15. Mersini-Houghton, Laura, *Before the Big Bang – The Origin of the Universe and What Lies Beyond,* 2022, p.70.
16. Sutter, Paul S., *Your Place in the Universe – Understanding Our Big Messy Existence*, 2018, p.86.

Chapter Three
The Origin of Fields

Particle Fields

1. Cham, Jorge & Daniel Whiteson, *We Have No Idea*, 2018, p.65.
2. Carroll, Sean, *The Big Picture – The Origins of Life, Meaning, and the Universe Itself*, 2017, p.173. Dr. Carroll tells us that, '…a field is something that stretches all throughout space, taking on some particular value at every point. Modern physics says

that the particles and the forces that make up all atoms all arise out of fields.'
3. Munowitz, Michael, *Knowing -- The Nature of Physical Law*, 2005, p.38.
4. Musser, George, *Spooky Action at a Distance – The Phenomenon that Reimagines Space and Time and What it Means for Black Holes, the Big Bang, and Theories of Everything*, 2015, p.134. Dr. Musser in his discussion of the development of quantum field theory, says that physicists combined elements of both quantum mechanics and relativity theory, performing a shotgun marriage with unforeseeable consequences. 'To this day, physicists struggle to fathom what quantum field theory is telling them about the world.' In the same paragraph he goes on to say with regard to quantum field theory, 'It has a deserved reputation as the most mathematically badass subject in the sciences. Even experts hang on for dear life.'
5. Munowitz, Michael, *Knowing -- The Nature of Physical Law*, 2005, p.322.
6. Carroll, Sean, *The Big Picture – The Origins of Life, Meaning, and the Universe Itself*, 2017, p.176.
7. Kaku, Michio, *The God Equation, The Quest for a Theory of Everything*, 2021, p.58. Dr. Kaku in his relating the history of discovery in particle physics from the perspective of what scientists believed to be true at the time, says, 'Light was made of photons, which are quanta, or particles, but each photon created fields around it (the electric and magnetic fields). These fields, in turn, were shaped like waves and obeyed Maxwell's equations. We now have a beautiful relationship between particles and the fields that surround it.' Although he was not arguing specifically that it is the fields that are doing the waving, his choice of words certainly suggests that is the case.
8. Ball, Philip, *Beyond Weird – Why Everything You Thought About Quantum Physics is Different*, 2018, p.169.
9. McTaggart, Lynne, *The Field – The Quest for the Secret Force of the Universe*, 2002, p.43.

A Particle's Personal Field
10. James, Tim, *Fundamental – How Quantum and Particle Physics Explain Absolutely Everything*, 2020, p.50.
11. Nomura, Yasunori & Bill Poirier & John Terning, *Quantum Physics, Mini Black Holes, and the Multiverse – Debunking Common Misconceptions in Theoretical Physics*, 2018, p.16.
12. Geach, James, *Five Photons – Remarkable Journeys of Light Across Space and Time*, 2018, p.73.

The Double-Slit Experiment

13. Ford, Kenneth W., *The Quantum World - Quantum Physics for Everyone*, 2005, p.185.
14. Ball, Philip, *Beyond Weird – Why Everything you Thought you knew about Quantum Physics is Different*, 2020, p.65-75.
15. Ford, Kenneth W., *The Quantum World - Quantum Physics for Everyone*, 2005, p.196.
16. Smolin, Lee, *Einstein's Unfinished Revolution – The Search for What Lies Beyond the Quantum*, 2019, p.100.
17. Becker, Adam, *What is Real -- The unfinished quest for the meaning of quantum physics*, 2019, p.89.
18. James, Tim, *Fundamental – How Quantum and Particle Physics Explain Absolutely Everything*, 2020, p.108. James in pointing out that de Broglie admitted he made a mistake introducing wave-particle duality and that electrons and protons were particles only. He goes on to say, 'They do not have wave character, but instead were surrounded by some background substance that did have waves in it. The particles were pushed around by these invisible "guide waves" and thus would appear to move in wave-like trajectories.' According to the Tor Model, de Broglie got it right and that the background substance is Penergy. Unfortunately, no one believed him.
19. Becker, Adam, *What is Real -- The unfinished quest for the meaning of quantum physics*, 2019, p.102.
20. Baggott, Jim, *Quantum Reality – The Quest for the Real Meaning of Quantum Mechanics – A Game of Theories*, 2020, p.181.
21. Ball, Philip, *Beyond Weird – Why Everything you Thought you knew about Quantum Physics is Different*, 2020, p.110.

Chapter Four
The Origin of Forces

1. Beckman, Milo, *Math Without Numbers*, 2021, p.191.
2. Kaku, Michio, *The God Equation, The Quest for a Theory of Everything*, 2021, p.101.

The Theoretical Basis for Force

3. Hawking, Stephen, *A Brief History of Time*, 1988, p.68-69.
4. Baggott, Jim, *Origins -- The Scientific Story of Creation*, 2015, p.79-80.
5. Butterworth, Jon, *Atom Land – A Guided Tour Through the Strange (and Impossibly Small) World of Particle Physics*, 2019, p.217.
6. Hawking, Stephen, *A Brief History of Time*, 1988, p.157.

NOTES & REFERENCES

7. Kaku, Michio, *The God Equation – The Quest for a Theory of Everything*, 2021, p.104. Dr. Kaku tells us that after scientists introduced the theory of the graviton that, 'Sadly, the bag of tricks painfully accumulated by physicists for the past seventy years to eliminate these infinities failed for the graviton. Here, physicists hit a brick wall.'
8. Ford, Kenneth W., *The Quantum World - Quantum Physics for Everyone*, 2005, p.86-87. In his discussion on bosons and the creation/annihilation of particles, he says, '...the interaction event is a truly catastrophic event in which every particle is either annihilated or created.' In referring to a Feynman diagram he says, 'At point A, the incoming electron is destroyed, a photon is created, and a *new* electron is created.' What he is saying overall is that every particle interaction creates a virtual boson and two new particles. This phenomenon was also noted in *We Have No Idea* by Jorge Cham and Daniel Whiteson, who say on page 212, 'That means, in a way, all particle interactions result in annihilation of the original particles into new particles."
9. Musser, George, *Spooky Action at a Distance – The Phenomenon that Reimagines Space and Time and What it Means for Black Holes, the Big Bang, and Theories of Everything*, 2015, p.73. Dr. Musser does not make this observation incidental to an argument for force being conveyed by fields rather than virtual particles. Nonetheless, if his observation is accurate, it could be used, as I am using it, to suggest that force showing a slight delay when conveyed, better suits the idea of it being conveyed through a medium such as Penergy, rather than by virtual boson exchange. A further description of the delay between forces can be found in Marc Lange's *Philosophy of Physics*, page 29.
10. Krauss, Lawrence, *The Greatest Story Ever Told – So Far - Why are We Here*, 2018, p.104. Dr. Krauss here is making the point that since a virtual particle's time in existence is inversely proportional to its energy, a photon carrying very little energy could travel as far as Alpha Centauri. My point is that given that is true, the electron on earth could also have the same relationship with every electron between earth and Alpha Centauri, making the constant exchange of photons ridiculous, or if a constant exchange is not taking place with all those electrons, there is nothing more than an intermittent force between them.

11. James, Tim, *Fundamental – How Quantum and Particle Physics Explain Absolutely Everything*, 2020, p.180-181.
12. Kaku, Michio, *The God Equation – The Quest for a Theory of Everything*, 2021, p.138.
13. Perlov, Delia, and Alex Vilenkin, *Cosmology for the Curious*, 2017, p.306.

The Magnetic Force
14. Schumm, Bruce A., *Deep Down Things -- The Breath-Taking Beauty of Particle Physics,* 2004, p.88.

The Gravitational Force
15. Impey, Chris, *Einstein's Monsters – The Life and Times of Black Holes,* 2019, p.174.
16. Clark Stuart, *The Unknown Universe – A New Exploration of Time, Space, and Modern Cosmology.* 2016, p.159.
17. Butterworth, Jon, *Atom Land – A Guided Tour Through the Strange (and Impossibly Small) World of Particle Physics*, 2019, p.106.

The Weak Force
18. Wilczek, Frank, *A Beautiful Question – Finding Nature's Deep Design*, 2015, p.260.
19. Krauss, Lawrence, *The Greatest Story Ever Told – So Far - Why are We Here,* 2018, p.253.

Chapter Five
The Origin of Composite Particles

The Cosmic Environment
1. Butterworth, Jon, *Atom Land – A Guided Tour Through the Strange (and Impossibly Small) World of Particle Physics,* 2019, p.120.
2. Schumm, Bruce A., *Deep Down Things -- The Breath-Taking Beauty of Particle Physics,* 2004, P.98.
3. Smolin, Lee, *The Trouble with Physics – The Rise of String Theory, the Fall of Science, and What Comes Next,* 2007, P.73.
4. Devereux, Carolyn, *Cosmological Clues – Evidence for the Big Bang, Dark Matter, and Dark Energy,* 2021, p.116.
5. Devereux, Carolyn, *Cosmological Clues – Evidence for the Big Bang, Dark Matter, and Dark Energy,* 2021, p.74.
6. Greenstein, George, *Symbiotic Universe – Life and Mind in the Cosmos,* 1988, p.164.
7. Hawking, Stephen & Leonard Mlodinow, *The Grand Design,* 2010, p.129.

8. Guth, Alan H., *The Inflationary Universe – The Quest for a New Theory of Cosmic Origins*, p.238.
9. James, Tim, *Astronomical -From Quarks to Quasars, the Science of Space at Its Strangest*, 2020, p.57.
10. Carroll, Sean, *The Big Picture – The Origins of Life, Meaning, and the Universe Itself*, 2017, p.308.
11. Davies, Paul, *The Cosmic Blueprint – New Discoveries in Nature's Creative Ability to Order the Universe*, 1989, p.5.

Cosmic Particle Attributes

12. Alexander, Stephon, *Fear of a Black Universe – An Outsiders Guide to the Future of Physics*, 2021, p.65.
13. Schumm, Bruce A., *Deep Down Things -- The Breath-Taking Beauty of Particle Physics*, 2004, P.186.
14. Schumm, Bruce A., *Deep Down Things -- The Breath-Taking Beauty of Particle Physics*, 2004, P.187.

The Origin of Particle Mass

15. Wilczek, Frank, *Fundamentals – Ten Keys to Reality*, 2021, p.57. Dr. Wilczek tells us that, 'The Higgs particles, whose discovery was a major triumph for twenty-first century physics, is highly unstable. It lives for only about 10^{-22} second. Thus, in order to discern evidence for its existence, physicists had to reconstruct events on that time scale.'
16. Dine, Michael, *This Way to the Universe*, 2022, p.138.
17. Krauss, Lawrence, *The Greatest Story Ever Told – So Far - Why are We Here*, 2018, p.271.
18. Kaku, Michio, *The God Equation – The Quest for a Theory of Everything*, 2021, p.89.
19. Padilla, Antonio, *Fantastic Numbers and Where to Find Them – A Cosmic Quest from Zero to Infinity*, 2022, P.190.
20. Hossenfelder, Sabine, *Lost in Math – How Beauty Leads Physics Astray*, 2020, p.37-38. Dr. Hossenfelder says, '...quantum fluctuations make a huge contribution to the Higgs mass. Contributions like this are normally small, but for the Higgs they lead to a mass much larger than what is observed – too large, indeed, by a factor of 10^{14}. Not a little bit off, but dramatically, inadmissibly wrong.' She goes on to explain how the numbers can be adjusted and points out that the final number requires an explanation, and that numbers requiring an explanation are called *fine-tuned*. She says, 'In the Standard Model, the Higgs mass is not natural, which makes it ugly.'
21. Still, Ben, *Particle Physics Brick by Brick – Atomic and Subatomic Physics Explained in Legos*, 2018, p.163.

22. Krauss, Lawrence, *The Greatest Story Ever Told – So Far - Why are We Here,* 2018, p.255-256. Dr. Krauss tells us, 'These particles that interact with the Higgs background field then experience a kind of resistance to their motion that slows their travel to less than the speed of light-just as a swimmer in molasses will move more slowly than a swimmer in water. Once they are moving at sub-light speed, the particles behave as if they are massive.'
23. Krauss, Lawrence, *The Greatest Story Ever Told – So Far - Why are We Here,* 2018, p.256. Here Dr. Krauss is pointing out that 'For every new field in nature, at least one new type of elementary particle must exist with that field.' He describes this as a central property of quantum field theory.
24. Fritzsch, Harald, *The Fundamental Constants – A Mystery of Physics*, 2005, p.113.
25. Lindley, David, *The Dream Universe -- How Fundamental Physics Lost its Way*, 2020, p.151.
26. Lincoln, Don, *Einstein's Unfinished Dream*, 2023, p.42.
27. Krauss, Lawrence, *The Edge of Knowledge -Unsolved Mysteries of the Cosmos,* 2023, p.88.
28. Impey, Chris, *Einstein's Monsters – The Life and Times of Blackholes,* 2019, p.174. Dr. Impey tells us, '…in Einstein's theory [of General Relativity], mass couple to the geometry of space-time. In 1918 in was predicted that the rotation of a massive object would distort space-time… This twisting of the contours of space is called frame dragging.' He goes on to relate the story of the experiment than confirmed this prediction, and how Einstein's prediction of space-time curvature and frame dragging was confirmed.
29. Wilczek, Frank, *Fundamentals – Ten Keys to Reality*, 2021, p.74.
30. Lange, Marc, *An Introduction to The Philosophy of Physics – Locality, Fields, Energy, Mass,* 2002, p.169.
31. Wilczek, Frank, *Fundamentals – Ten Keys to Reality,* 2021, p.74.
32. Still, Ben, *Particle Physics Brick by Brick – Atomic and Subatomic Physics Explained in Legos,* 2018, p.117.
33. Smith, Timothy Paul*, Hidden Worlds – Hunting for Quarks in Ordinary Matter,* 2003, p.63.
34. Padilla, Antonio, *Fantastic Numbers and Where to Find Them – A Cosmic Quest from Zero to Infinity,* 2022, p.189*.*

Virtues of Composite Elementary Particles
35. Fritzsch, Harald, *The Fundamental Constants – A Mystery of Physics*, 2005, p.27. Dr. Fritzsch points out that treating the electron as point-like leads to nonsensical infinities, but if the

NOTES & REFERENCES

electron is considered to have a radius, the infinities go away and the radius appears in their place in the calculations. This seems like a good argument for electrons not being infinitely small, point-like particles.

36. Fritzsch, Harald, *The Fundamental Constants – A Mystery of Physics*, 2005, p.116.
37. Barrow, John D. and Joseph Silk, *The Left Hand of Creation -- The Origin and Evolution of the Expanding Universe*, 1993, p.142.
38. Lindley, David, *The Dream Universe -- How Fundamental Physics Lost its Way*, 2020, p.136.
39. Lincoln, Don, *Einstein's Unfinished Dream*, 2023, p.239.
40. Butterworth, Jon, *Atom Land – A Guided Tour Through the Strange (and Impossibly Small) World of Particle Physics*, 2019, p.275. Dr. Butterworth speculates on the prospect of the quarks, leptons, and bosons of the Standard Model may contain smaller constituents, just as the atoms on the Periodic Table turned out to be made of other things.
41. Ford, Kenneth W., *The Quantum World - Quantum Physics for Everyone*, 2004, p.61.
42. Munowitz, Michael, *Knowing -- The Nature of Physical Law*, 2005, p.34.
43. Hawking, Stephen, *A Brief History of Time*, 1988, p.167.
44. Cham, Jorge & Daniel Whiteson, *We Have No Idea – A Guide to the Unknown Universe*, 2018, p.50.
45. Impey, Chris, *How it Began – A Time Traveler's Guide to the Universe*, 2012, p.298.
46. Cham, Jorge & Daniel Whiteson, *We Have No Idea – A Guide to the Unknown Universe*, 2018, p.54.

Chapter Six
The Origin of the Atom

Quark Configurations
1. Butterworth, Jon, *Atom Land – A Guided Tour Through the Strange (and Impossibly Small) World of Particle Physics*, 2019, p.137.

Meson Configurations
2. Still, Ben, *Particle Physics Brick by Brick – Atomic and Subatomic Physics Explained in Legos*, 2018, p.121.

Electron Configuration
3. Wilczek, Frank, *Fundamentals – Ten Keys to Reality*, 2021, p.89.
4. Alexander, Stephon, *Fear of a Black Universe – An Outsiders Guide to the Future of Physics*, 2021, p.70.

Photon Configuration
5. Baggott, Jim, *Origins – The Scientific Story of Creation*, 2015, p.45.

Evolution of a Nucleus
6. Perlov, Delia, and Alex Vilenkin, *Cosmology for the Curious*, 2017, p.189-190.
7. Baggott, Jim, *Origins – The Scientific Story of Creation*, 2015, p.66.
8. Baggott, Jim, *Origins – The Scientific Story of Creation*, 2015, p.64.

The CMB
9. Perlov, Delia, and Alex Vilenkin, *Cosmology for the Curious*, 2017, p.232.

Chapter Seven
Speculations on Unsolved Mysteries

Speculations on the Origin of Dark Matter
1. Davies, Paul and John Gribbin, *The Matter Myth - Dramatic Discoveries that Challenge our Understanding of Physical Reality*, 1992, p.176.
2. Alexander, Stephon, *Fear of a Black Universe – An Outsiders Guide to the Future of Physics*, 2021, p.113.

Speculations on the Origin of Dynamic Gravity
3. Clark Stuart, *The Unknown Universe – A New Exploration of Time, Space, and Modern Cosmology*, 2016, p.162.
4. Impey, Chris, *Einstein's Monsters – The Life and Times of Black Holes*, 2019, p.174. Dr. Impey says, 'Gravity does not depend on rotation in Newton's theory. But in Einstein's theory, mass couples to space-time. In 1918 it was predicted that the rotation of a massive object would distort space-time, making the orbit of a smaller nearby object precess, like the pivoting of a spinning top. This twisting of the contours of space is called frame-dragging.'
5. Hawking, Stephen, *Black Holes and Baby Universes, and Other Essays*, 1994, p.75.
6. Nomura, Yasunori & Bill Poirier & John Terning, *Quantum Physics, Mini Black Holes, and the Multiverse – Debunking Common Misconceptions in Theoretical Physics*, 2018, p.97.
7. Impey, Chris, *Einstein's Monsters – The Life and Times of Black Holes*, 2019, p.176.
8. Rezzolla, Luciano, *The Irresistible Attraction to Gravity*, 2023, p.98.

NOTES & REFERENCES

9. Devereux, Carolyn, *Cosmological Clues*, 2021, p.91. Dr. Devereux in her discussion on the topic of MOND, an alternative accounting for galaxy rotation without dark matter, she says, 'Mond not only explains the rotation curves without the need for dark matter. It does more. From the equations it can be shown that the mass of a galaxy is related to its velocity of rotation (v) to the power (v^4).' This observation is in keeping with the Tor Model's theory of blackhole mass being related in part to the blackhole's spin rate.

Speculations on Particle Creation and Decay

10. Munowitz, Michael, *Knowing -- The Nature of Physical Law*, 2005, p.32.
11. www.brittanica.com/science/beta-decay. This point is also expressed in Paul Davies' *What's Eating the Universe*, who tells us that electrons move at near light speed during radioactive emission.
12. Ford, Kenneth W., *The Quantum World - Quantum Physics for Everyone*, 2005, p.86.

Speculations on Why Space & Time are Relative

13. Krauss, Lawrence, *The Greatest Story Ever Told – So Far - Why are We Here*, 2018, p.81. Dr. Krauss was comparing his idea that the universe popped into existence from nothing to the production of photons created from atoms. He says, "Electrons in hot atoms emit photons – photons that didn't exist until they were emitted – which are emitted spontaneously and without specific cause." Using the notion that photons don't exist until emitted and don't exhibit any mass, then they can't have had any momentum or energy that would be utilized in the calculation for their motion with respect to the train.
14. Cowen, Ron, *Gravity's Century – From Einstein's Eclipse to Images of Black Holes*, p.93.
15. Clegg, Brian, *Gravity*, 2016, p.128.
16. Munowitz, Michael, *Knowing -- The Nature of Physical Law*, 2005, p.106.
17. Clegg, Brian, *Gravity*, 2016, p.174.
18. Close, Frank, *Antimatter*, 2018, p.28.
19. Gribbin, John, *In the Beginning – After COBE and Before the Big Bang*, 1993, p.223.
20. Turner, Ben, *Time Moved Five Times Slower*, Live Science Magazine, July 2023.
21. Lindley, David, *The Dream Universe -- How Fundamental Physics Lost its Way*, 2020, p.160.

Mathematics and Uncertainty

22. Hossenfelder, Sabine, *Existential Physics – A Scientist's Guide to Life's Biggest Questions*, 2022, p.20.

Chapter Eight
The Origin of Reality

Finding Reality

1. Smolin, Lee, *The Trouble with Physics – The Rise of String Theory, the Fall of Science, and What Comes Next*, 2007, p.7.
2. Wilczek, Frank, *A Beautiful Question – Finding Nature's Deep Design*, 2015, p.9.
3. Kaku, Michio, *Visions- How Science will Revolutionize the 21st Century*, 1997, p.109.

Uncertainty

4. Nomura, Yasunori & Bill Poirier & John Terning, *Quantum Physics, Mini Black Holes, and the Multiverse – Debunking Common Misconceptions in Theoretical Physics*, 2018, p.21.
5. Krauss, Lawrence, *The Greatest Story Ever Told – So Far - Why are We Here*, 2018, p.91.
6. Smolin, Lee, *The Trouble with Physics – The Rise of String Theory, the Fall of Science, and What Comes Next*, 2007, p.7.
7. Nomura, Yasunori & Bill Poirier & John Terning, *Quantum Physics, Mini Black Holes, and the Multiverse – Debunking Common Misconceptions in Theoretical Physics*, 2018, p.21.
8. Nomura, Yasunori & Bill Poirier & John Terning, *Quantum Physics, Mini Black Holes, and the Multiverse – Debunking Common Misconceptions in Theoretical Physics*, 2018, p.21.
9. Fritzsch, Harald, *The Fundamental Constants – A Mystery of Physics*, 2005, p.26.
10. Huang, Kerson, *Fundamental Forces of Nature -- The Story of Gauge Fields*, 2007, p.97.
11. Carroll, Sean, *The Big Picture – The Origins of Life, Meaning, and the Universe Itself*, 2017, p.177.

Particle Entanglement

12. Becker, Adam, *What is Real -- The unfinished quest for the meaning of quantum physics*, 2019, p.150.
13. Nomura, Yasunori & Bill Poirier & John Terning, *Quantum Physics, Mini Black Holes, and the Multiverse – Debunking Common Misconceptions in Theoretical Physics*, 2018, p.63.
14. Clegg, Brian, The God Effect – *Quantum Entanglement, Science's Strangest Phenomenon*, 2006, p.34.

15. Becker, Adam, What is Real -- *The unfinished quest for the meaning of quantum physics,* 2019, p.150-151.
16. Nomura, Yasunori & Bill Poirier & John Terning, *Quantum Physics, Mini Black Holes, and the Multiverse – Debunking Common Misconceptions in Theoretical Physics,* 2018, p.22.

The Wavefunction

17. Smolin, Lee, *Einstein's Unfinished Revolution – The Search for What Lies Beyond the Quantum,* 2019, p.128.
18. Nomura, Yasunori & Bill Poirier & John Terning, *Quantum Physics, Mini Black Holes, and the Multiverse – Debunking Common Misconceptions in Theoretical Physics,* 2018, p.64.
19. Nomura, Yasunori & Bill Poirier & John Terning, *Quantum Physics, Mini Black Holes, and the Multiverse – Debunking Common Misconceptions in Theoretical Physics,* 2018, p.38.
20. Carroll, Sean, *Something Deeply Hidden – Quantum Worlds and the Emergence of Spacetime,* 2020, p.20.
21. Ball, Philip, *Beyond Weird – Why Everything you Thought you knew about Quantum Physics is Different,* 2020, p.114.
22. Becker, Adam, *What is Real -- The unfinished quest for the meaning of quantum physics,* 2019, p.17.

The Reality of Quantum Theory

23. Carroll, Sean, *Something Deeply Hidden – Quantum Worlds and the Emergence of Spacetime,* 2020, p.18.
24. Nomura, Yasunori & Bill Poirier & John Terning, *Quantum Physics, Mini Black Holes, and the Multiverse – Debunking Common Misconceptions in Theoretical Physics,* 2018, p.63. Dr. Nomura says, '…Copenhagen [interpretation] does make certain metaphysical claims. One of these is that wavefunction collapse occurs as a result of measurement. Another is that physical systems do not actually possess attributes until those attributes are measured.' Her point is that neither of those positions are backed by scientific evidence.
25. Kaku, Michio, *Visions- How Science will Revolutionize the 21st Century,* 1997, p.351.
26. Becker, Adam, *What is Real -- The unfinished quest for the meaning of quantum physics,* 2019, p.97.
27. Rovelli, Carlo, *Reality is Not What it Seems -- The Journey to Quantum Gravity,* 2017, p.178. Dr. Rovelli devotes an entire chapter to time, entitled Time Does Not Exist.
28. Becker, Adam, *What is Real -- The unfinished quest for the meaning of quantum physics,* 2019, p.5. Dr. Becker in explaining Bohr's interpretation of quantum mechanics says, 'To

them, the theory needs no interpretation, because the things that the theory describes aren't truly real. Indeed, the strangeness of quantum phenomena has led some prominent physicists to state flatly that there is no alternative, that quantum physics proves that small objects simply do not exist is the same objectively real way as the objects in our everyday lives do.'

29. Rovelli, Carlo, *Reality is Not What it Seems -- The Journey to Quantum Gravity*, 2017, p.129. Dr. Rovelli says, 'The world is not made up of fields and particles but a single type of entity: the quantum field. There are no longer particles that move in space with the passage of time, but quantum fields whose elementary events happen in spacetime.'

30. Carroll, Sean, *Something Deeply Hidden – Quantum Worlds and the Emergence of Spacetime*, 2020, p.38-40.

31. James, Tim, Fundamental – *How Quantum and Particle Physics Explain Absolutely Everything*, 2020, p.47.

32. Carroll, Sean, *The Big Picture – The Origins of Life, Meaning, and the Universe Itself*, 2017, p.131-132. Dr. Carroll says math and science are completely different endeavors. 'Math is all about proving things, but the things that math proves are not true facts about the actual world. They are implications of various assumptions.'

33. Carroll, Sean, *The Big Picture – The Origins of Life, Meaning, and the Universe Itself*, 2017, p.35. Dr. Carroll says, 'Quantum mechanics has supplanted classical mechanics as the best way to talk about the universe at a deep level. Unfortunately, and to the chagrin of physicists everywhere, we don't fully understand what the theory actually is.'

34. Smolin, Lee, *Einstein's Unfinished Revolution – The Search for What Lies Beyond the Quantum*, 2019, p.29.

35. Ford, Kenneth W., *The Quantum World - Quantum Physics for Everyone*, 2005, p.128.

36. Nomura, Yasunori & Bill Poirier & John Terning, *Quantum Physics, Mini Black Holes, and the Multiverse – Debunking Common Misconceptions in Theoretical Physics*, 2018, p.3.

A Theory of Everything

37. Randall, Lisa, *Knocking on Heaven's Door – How Physics and Scientific Thinking Illuminate the Universe and the Modern World*, 2012, p.253.

38. Cliff, Harry, *How to Make an Apple Pie from Scratch*, 2021, p.339.

NOTES & REFERENCES

Chapter Nine
New Perspectives

The CAGI Perspective
1. Conover, Robert J., *Journey of the Universe – A New Perspective on its Past, Present, and Future Evolution,* 2021.

The Tor Model Perspective
2. Devereux, Carolyn, *Cosmological Clues – Evidence for the Big Bang, Dark Matter, and Dark Energy,* 2021, p.35.
3. Clark Stuart, *The Unknown Universe – A New Exploration of Time, Space, and Modern Cosmology,* 2016, p.250.

The Mathematical Perspective
4. Kaku, Michio, *The God Equation, The Quest for a Theory of Everything,* 2021, p.173. Dr. Kaku says, 'Any theory of everything will have an infinite number of solutions depending on the initial conditions. But how do you determine the initial conditions of the entire universe? This means you have to input the conditions of the Big Bang from the outside, by hand.' My point in this section is that the more conditions one inputs by hand, the greater the probability of initiating a pathway leading to false theories and realities.
5. Perlov, Delia, and Alex Vilenkin, *Cosmology for the Curious,* 2017, p.301-302.
6. Hawking, Stephen & Leonard Mlodinow, *The Grand Design,* 2010, p.162.
7. Baggott, Jim, *Origins – The Scientific Story of Creation,* 2015, p.4.

Cosmic Reality
8. Krauss, Lawrence M., *A Universe from Nothing – Why There is Something Rather than Nothing,* 2012, p.89
9. Rezzolla, Luciano, *The Irresistible Attraction to Gravity – A Journey to Discover Black Holes,* 2023, p.199.
10. Lincoln, Don, *Einstein's Unfinished Dream,* New York, 2023, p.280.

Bibliography

Alexander, Stephon, Fear of a Black Universe – An Outsiders Guide to the Future of Physics, Basic Books, New York, 2021.

Baggott, Jim, Origins -- The Scientific Story of Creation, Oxford University Press, 2015.

Baggott, Jim, Quantum Reality – The Quest for the Real Meaning of Quantum Mechanics – A Game of Theories, Oxford University Press, New York, 2020.

Ball, Philip, Beyond Weird – Why Everything You Thought About Quantum Physics is Different, University of Chicago Press, Chicago, IL, 2018.

Barnes, Luke A. & Geraint F. Lewis, The Cosmic Revolutionary's Handbook – Or: How to Beat the Big Bang, Cambridge University, 2020.

Barrow, John D. and Joseph Silk, The Left Hand of Creation -- The Origin and Evolution of the Expanding Universe, Oxford University Press, New York, 1993.

Becker, Adam, What is Real -- The Unfinished Quest for the Meaning of Quantum Physics, Basics Books, New York, 2019.

Beckman, Milo, Math Without Numbers, Dutton/Penguin Random House LLC, New York, 2021.

Butterworth, Jon, Atom Land – A Guided Tour Through the Strange (and Impossibly Small) World of Particle Physics, First Paperback Edition, The Experiment, New York, 2019.

Carroll, Sean, The Biggest Ideas in the Universe, Dutton – Penguin Random House, New York, 2022.

Carroll, Sean, The Big Picture – The Origins of Life, Meaning, and the Universe Itself, Dutton, an imprint of Penguin Random House, LLC, New York, Paperback Edition, 2017.

Carroll, Sean, Something Deeply Hidden – Quantum Worlds and the Emergence of Spacetime, Dutton, 2019. A good read for understanding Quantum Mechanics, the arguments for a Wave Function, Spacetime, Quantum Field Theory, Entanglement, and Multi-verses.

Carroll, Sean, Something Deeply Hidden – Quantum Worlds and the Emergence of Spacetime, Dutton, paperback edition, 2020.

Chaisson, Eric J., Cosmic Evolution – The Rise of Complexity in Nature, Harvard University Press, Cambridge, MA, 2001.

BIBLIOGRAPHY

Cham, Jorge & Daniel Whiteson, We Have No Idea – A Guide to the Unknown Universe, Riverhead Books, New York, Paperback Edition, 2018. An easy read, and a witty but professional look at what we don't know about the universe.

Clark Stuart, The Unknown Universe – A New Exploration of Time, Space, and Modern Cosmology, Pegasus Books, New York, 2016.

Clegg, Brian, Gravity, Duckworth Overlook, paperback edition, 2016.

Clegg, Brian, The God Effect – Quantum Entanglement, Science's Strangest Phenomenon, St. Martin's Press, New York, 2006. An excellent summary of the discoveries and theories behind some of the esoteric aspects of quantum physics.

Cliff, Harry, How to Make an Apple Pie from Scratch, Double Day, New York, 2021.

Close, Frank, Antimatter, Oxford University Press, United Kingdom, second edition paperback, 2018.

Close, Frank, The Infinity Puzzle – Quantum Field Theory and the Hunt for an Orderly Universe, Basic Books, New York, 2011.

Cowen, Ron, Gravity's Century – From Einstein's Eclipse to Images of Black Holes, Harvard University Press, Cambridge, Massachusetts, 2019.

Davies, Paul, The Cosmic Blueprint – New Discoveries in Nature's Creative Ability to Order the Universe, Simon & Schuster, New York, Touchstone Edition, 1989.

Davies, Paul and John Gribbin, The Matter Myth - Dramatic Discoveries that Challenge our Understanding of Physical Reality, Simon & Schuster, New York, 1992.

Devereux, Carolyn, Cosmological Clues – Evidence for the Big Bang, Dark Matter, and Dark Energy, CRC Press, Taylor & Francis Group, Baca Raton, Florida, 2021.

Dine, Michael, This Way to the Universe, Dutton/Penguin Random House, LLC, 2022.

Enos, Roland, The Science of Spin, Scribner/Simon & Schuster, New York, NY, 2023.

Farmelo, Graham, The Universe Speaks in Numbers – How Modern Math Reveals Nature's Deepest Secrets, Basic Books, Hachette Book Group, New York, NY, 2019.

Ford, Kenneth W., The Quantum World - Quantum Physics for Everyone, Cambridge, Mass.: Harvard University Press, Copyright © 2004, 2005 by the President and Fellows of Harvard College.

Fritzsch, Harald, The Fundamental Constants – A Mystery of Physics, World Scientific, Singapore, 2005.

Geach, James, Five Photons – Remarkable Journeys of Light Across Space and Time, Reaktion Books, London, 2018.

Gott, Richard, Cosmic Web – Mysterious Architecture of the Universe, Princeton University Press, Princeton, NJ Paperback Edition, 2018.

Gould, Roy R., Universe in Creation – A New Understanding of the Big Bang and the Emergence Life, Harvard University Press, London, 2018.

Greene, Brian, Until the End of Time – Mind, Matter, and our Search for Meaning in an Evolving Universe, Alfred A Knopf, New York, 2020.

Greenstein, George, Symbiotic Universe – Life and Mind in the Cosmos, William Morrow and Co., New York, 1988.

Gribbin, John, In the Beginning – After COBE and Before the Big Bang, Little, Brown and Co., New York, 1993.

Guth, Alan H., The Inflationary Universe – The Quest for a New Theory of Cosmic Origins, Perseus Books, Reading, Massachusetts, 1997. The story of the reasons for and the development of the inflation theory.

Han, M Y, Quarks and Gluons – A Century of Particle Charges, World Scientific, Singapore, 1999.

Hawking, Stephen, Black Holes and Baby Universes, and Other Essays, Bantam Books, New York, paperback Edition, 1994.

Hawking, Stephen & Leonard Mlodinow, The Grand Design, Bantam Books, New York, 2010.

Hawking, Stephen, A Brief History of Time - From the Big Bang to Black Holes, Bantam Books, New York, 1990.

Hawking, Stephen, The Theory of Everything – The Origin and Fate of the Universe, Jaico Publishing House, 2006.

Impey, Chris, Einstein's Monsters – The Life and Times of Blackholes, W.W. Norton & Co., New York, 2019.

Impey, Chris, How It Began – A Time Traveler's Guide to the Universe, W.W. Norton & Co., New York, paperback Edition, 2013.

Gregensen, Erik, Beta-Decay, Encyclopedia Britannica, www.britannica.com/science/beta-decay, April 21, 2022.

Hossenfelder, Sabine, Lost in Math – How Beauty Leads Physics Astray, Basic Books, Paperback Edition, New York, NY, 2020.

Hossenfelder, Sabine, Existential Physics – A Scientist's Guide to Life's Biggest Questions, Viking, New York, NY, 2022.

BIBLIOGRAPHY

Huang, Kerson, Fundamental Forces of Nature -- The Story of Gauge Fields, World Scientific Publishing Co., Singapore, 2007. A good summary of particle physics and the development of the gauge fields.

James, Tim, Astronomical -From Quarks to Quasars, the Science of Space at Its Strangest, Robinson/ Little, Brown Book Group, London, 2020.

James, Tim, Fundamental – How Quantum and Particle Physics Explain Absolutely Everything, Pegasus Books, New York, 2020.

Jones, Mark, Robert J.A. Lambourne, and Stephen Serjeant, Editors, An Introduction to Galaxies and Cosmology, Second Edition, Cambridge University Press, UK, 2015.

Kaku, Michio, The God Equation – The Quest for a Theory of Everything, Doubleday, New York, 2021.

Kaku, Michio, Physics of the Impossible – A Scientific Exploration into the World of Phasers, Force Fields, Teleportation, and Time Travel, Double Day, New York, 2008.

Kaku, Michio, Visions- How Science will Revolutionize the 21st Century, Double Day, New York, 1997.

Kisslinger, Leonard S., Astrophysics and The Evolution of the Universe, World Scientific, Singapore, 2017.

Krauss, Lawrence, The Greatest Story Ever Told – So Far - Why are We Here, Simon & Schuster, UK, Paperback Edition, 2018.

Krauss, Lawrence M., A Universe from Nothing – Why There is Something Rather than Nothing, Free Press – a Division of Simon & Schuster, New York, 2012.

Krauss, Lawrence, The Edge of Knowledge -Unsolved Mysteries of the Cosmos, Post Hill Press, New York, NY, 2023.

Lange, Marc, An Introduction to The Philosophy of Physics – Locality, Fields, Energy, Mass, Blackwell Publishing, UK, 2002.

Lincoln, Don, Einstein's Unfinished Dream, Oxford University Press, New York, 2023.

Lindley, David, The Dream Universe -- How Fundamental Physics Lost its Way, Doubleday, New York, 2020. An easy read recounting the problems with our current theories on how the universe works.

Livio, Mario, Accelerating Universe – Infinite Expansion, The Cosmological Constant, and the Beauty of the Cosmos, John Wiley & Sons, Inc., New York and Canada, 2000.

Long, Ariana S., Ancient Galaxy Clusters Offer Clues About the Early Universe, Scientific American, Dec. 21, 2021.

McTaggart, Lynne, The Field – The Quest for the Secret Force of the Universe, Harper Collins Publishers, New York, 2002. An excellent story of the investigation into the mind, consciousness, and the Zero Point Field.

Mersini-Houghton, Laura, Before the Big Bang – The Origin of the Universe and What Lies Beyond, Mariner Books, Harper Collins Publishers, New York, 2022.

Munowitz, Michael, Knowing -- The Nature of Physical Law, Oxford University Press, New York, 2005.

Musser, George, Spooky Action at a Distance – The Phenomenon that Reimagines Space and Time and What it Means for Black Holes, the Big Bang, and Theories of Everything, Scientific American/Farrer, Strauss and Giroux, 2015.

Nomura, Yasunori & Bill Poirier & John Terning, Quantum Physics, Mini Black Holes, and the Multiverse – Debunking Common Misconceptions in Theoretical Physics, Multiversal Journeys, Springer International, Switzerland, 2018.

Padilla, Antonio, Fantastic Numbers and Where to Find Them – A Cosmic Quest from Zero to Infinity, Farrar, Straus and Giroux, New York, 2022.

Perlov, Delia, and Alex Vilenkin, Cosmology for the Curious, Springer International Publishing AG, 2017.

Quinn, Helen R. & Yossi Nir, The Mystery of the Missing Anti-matter, Princeton University Press, New Jersey, 2008. A good account of the several arguments regarding why there is only matter in the universe.

Randall, Lisa, Dark Matter and the Dinosaurs – The Astonishing Interconnectedness of the Universe, Harper Collins Publishers, New York, 2015.

Randall, Lisa, Knocking on Heaven's Door – How Physics and Scientific Thinking Illuminate the Universe and the Modern World, Harper Collins Publishers, New York, paperback edition, 2012.

Rezzolla, Luciano, The Irresistible Attraction to Gravity – A Journey to Discover Black Holes, Cambridge University Press, UK, 2023.

Rothman, Tony, A Little Book About the Big Bang, Harvard University Press, Cambridge, Mass., 2022.

Rovelli, Carlo, Reality is Not What it Seems -- The Journey to Quantum Gravity, Riverhead Books, New York, 2017. Good examination of spacetime, reality, relativity, quantum mechanics, and quantum fields.

BIBLIOGRAPHY

Schumm, Bruce A., Deep Down Things -- The Breath-Taking Beauty of Particle Physics, John Hopkins University Press, Baltimore, 2004.

Seife, Charles, Alpha & Omega – The Search for the Beginning and End of the Universe, Viking/Penguin Group, New York, 2003.

Siegel, Ethan, What Rules the Proton: Quarks or Gluons?, Forbes Magazine, 2021.

Smethurst, Becky, A Brief History of Blackholes – And Why Nearly Everything You Know About Them is Wrong, Macmillan Publishing, London, 2022.

Smith, Timothy Paul, Hidden Worlds – Hunting for Quarks in Ordinary Matter, Princeton University Press, Princeton and Oxford, 2003.

Smolin, Lee, Einstein's Unfinished Revolution – The Search for What Lies Beyond the Quantum, Penguin Press, New York, 2019.

Smolin, Lee, The Trouble with Physics – The Rise of String Theory, the Fall of Science, and What Comes Next, Houton Miffin Co., First Mariner Books Edition, New York, 2007.

Sokol, Joshua, Earliest Black Hole Gives Rare Glimpse of Ancient Universe, Quanta Magazine, Dec. 6, 2017.

Still, Ben, Particle Physics Brick by Brick – Atomic and Subatomic Physics Explained in Legos, Firefly Books, Ltd., New York, 2018.

Sutter, Paul S., Your Place in the Universe – Understanding Our Big Messy Existence, Prometheus Books, New York, 2018.

Tyson, Neil deGrasse and Donals Goldsmith, Origins, -- Fourteen Billion Years of Cosmic Evolution, W.W. Norton & Co, New York, 2005, re-issued 2014.

Tucker, Wallace H., Chandra's Cosmos, Smithsonian Books, Washington D.C., 2017.

Turner, Ben, Time moved '5 times slower' in the early universe, mind-bending black hole study reveals, Live Science Magazine, July 2023.

Wilczek, Frank, A Beautiful Question – Finding Nature's Deep Design, Penguin Books, 2015.

Wilczek, Frank, Fundamentals – Ten Keys to Reality, Penguin Press, New York, 2021.

INDEX

A

accretion, 80
Alexander, Stephon, 101
alpha decay, 153
Andromeda galaxy, 17
angular momentum, 24, 104, 110
annihilation
    avoidance, 38
anthropic principle, 210
anti-matter
    integral part of matter, 37
    mystery of, 36
atom
    evolution of, 138

B

Baggott, Jim, 211
Ball, Philip, 4, 51, 57, 60
Barnes, Luke, 22
Barrow, John, 112
Beckman, Milo, 63
Bell, John, 183
beta decay, 153
big bang, 6, 8, 19, 40
Big Bang Theory, 14
binding energy, 103, 130
blackhole
    accretion, 14
    affect on time passage, 161
    creation theory, 12
    made from pure energy, 19
    mergers of, 14
    super massive, 82

Blackhole. *See* Supermassive Blackhole
Bohm, David, 59, 174
Bohr, Niels, 171, 173
boneyard
    for particle creation, 152
boson
    configuration, 131
Butterworth, Jon, 91, 113

C

CAGI, 100, 128, 192, 201, 211
Carroll, Sean, 49, 96, 186, 188
Cascade Effect
    and homogeneity, 96
    evidence for, 21
    illustration of, 18
Casmir Affect, 65
centripetal force, 81
Cham, Jorge, 42, 113, 115
charge
    definition of, 76, 123
chemical energy
    definition of, 197
Clark, Stuart, 80, 148, 207
Clegg, Brian, 157
CMB, 140
cold trap, 39
Color Charge, 83
Comb. & Growth Imperative, 99
complexity, 31, 100
composite particles, 112
    arguments for, 114
conduction energy
    definition of, 196
convection energy

247

INDEX

definition of, 196
Cooper Pair, 131
Copenhagen Interpretation, 173
cosmic environment, 94
cosmic evolution, 98
   first phase of, 47
Cosmic Microwave Background, 97, 140
cosmic rays, 181
cosmic reality, 212
cosmic recycling, 99
cosmic web
   depiction of, 26
cosmological constant problem, 206
Cosmological Principle, 95

D

Dark Energy
   origin of, 40
dark matter, 151
   definition of, 104
   its origin, 145
Davies, Paul, 99, 146
de Broglie, Louis, 59, 60, 172, 174
de Coulomb, Augustin, 73
density fluctuations, 142
Devereux, Carolyn, 5, 94, 203
double-slit experiment, 56, 182
down quark
   decay, 87
   decay requiring a boson, 154
   possible configuration of, 125
drag the surrounding spacetime, 148
dynamic gravity, 150
   origin of, 148

E

$E=MC^2$, 2, 8, 106, 200
Early Elementary Particles, 20
   attributes of, 34
   origin of, 30
earthbound Sixth Level of matter, 210
Einstein, 6, 43, 159, 171, 173
   & the wave function, 187
   field equations, 13
   particle entanglement, 183
electric charge
   definition of, 76
   the value of, 123
   Tor count net differential, 123
electric force
   & matter/anti-matter attraction, 73
electromagnetic force, 76
electron
   configuration, 129
   illustration of, 130
electron volt
   definition of, 106
electroweak theory, 108
elliptical galaxy, 16
emergent qualities, 101, 192, 208, 210, 212
   of atoms, 139
   of Tryks, 117
entanglement. *See* Particle Entanglement
entropy, 100
event horizon, 12
Everett, Hugh, 174
evolution
   progresses in Levels, 98

F

Faraday, Michael, 73
field, 35
   local, 50
field energy, 198
fine-tuning problem, 210
first particles
   creation of, 8
first stars., 27
flexible connector, 119
force
   per Standard Model, 64
   per Tor Model, 68
   virtual particle exchange, 64
forced decay
   definition of, 153
Ford, Kenneth, 66, 113, 190
Fractional Quantum Hall Effect, 129
frame dragging, 42, 161
Franklin, Ben, 72
Friedmann, Alexander, 5
Fritzsch, Harald, 108, 112, 180

G

gamma rays, 181
Geach, James, 54
Gell-Mann Murry, 83
General Relativity, 43
Gluons
   mediator of strong force, 83
goal of physics, 190
gravitational condensing
   & time dilation, 160
gravitational energy, 198
gravitational field
   alternative theory for, 79
gravitational force, 78
gravitational mass, 80
gravitational waves, 42, 181
graviton
   and gravitational force, 78
   force carrying boson, 65
gravity, 149
Gribbin, John, 9, 23, 146
Guth, Alan, 95

H

Hawking, Stephen, 5, 13, 95, 113, 149, 211
Heisenberg Uncertainty Principle, 175
Heisenberg, Werner, 175
hidden variables, 51, 183
Higgs boson, 106
Higgs field
   gives particles mass, 107
   how created, 106
Higgs, Peter, 106
high speed compression
   & time dilation, 162
homogeneity problem, 94
Hooke, Robert, 172
horizon problem, 94, 205
Hossenfelder, Savine, 167
Hot Big Bang Theory, 7, 14, 89
Huang, Kerson, 180
Hubble Space Telescope, 14
Hubble, Edwin, 40
Huygens, Christian, 172

I

Impey, Chris, 5, 23, 24, 113
inertial mass, 80
inflation
   definition of, 95
initial conditions, 209
interspatial medium

argument for, 42
intrinsic angular momentum, 104
inverse square law
   the definition of, 78

J

James Webb Space Telescope, 14
James, Tim, 53, 67, 96, 188
Journey of the Universe, 202
JWST, 19, 22

K

Kaku, Michio, 6, 39, 50, 63, 66, 67, 171, 209
Krauss, Lawrence, 107, 213

L

Large Hadron Collider, 83, 87, 107, 118
Lewis, Geraint, 22
Lincoln, Don, 108, 113, 215
Lindley, David, 108, 112, 164
Livio, Mario, 41
Logical Positivism, 173

M

magnetic field, 70
magnetic force, 69
magnetism and electricity
   relationship between, 135
Majorana, 120
many worlds, 188
Many-Worlds Theory, 174
mass
   alternative theory for, 109
   and particle configuration, 120
mass of a blackhole, 14

matter and *anti-matter*, 35
McTaggart, Lynne, 52
Mersini-Houghton, Laura, 43
meson, 38
   configuration, 128
micro-macro world
   dividing line, 51
Milgrom, Mordehai, 150
minute environment
   of a particle, 190
mirror image, 35
Modified Newtonian Dynamics, 151
Munowitz, Michael, 49, 113, 154
Musser, George, 49, 66

N

natural selection, 92, 93, 114, 127
neutrino, 180
   configuration, 121
   oscillation, 120
neutron decay, 87
Newton, Issac, 165, 172
nexus
   between mass & gravity, 80
NIB-state, 11, 13, 40, 149
   definition of, 6
Nomura, Yasunori, 187, 190
now
   no such thing as, 178
nucleus evolution
   per Standard Model, 136
   per Tor Model, 137

O

Occam's razor, 126
of particle fragments, 147
orientational inertia, 110
origin of the CMB, 141

INDEX

P

Padilla, Antonio, 111
particle
  angular momentum, 104
  annihilation, 35
  complexity, 31
  consistency, 103
  creation, 151
  decay, 153
  few primary traits, 39
  generations of, 33
  inbalance of, 36
  never seen as waves, 53
  resilence, 34
  spin signature, 38
  tumble, 105
  virtual, 41, 64
particle evolution
  sequence of, 139
particle fields
  attributes of, 50
  field collapse, 58
  in Tor Model, 53
  origin of, 48
  why QM works, 55
particle mass, 106
particle minute environment, 181
particle pair, 32, 152
particle tunneling, 56
Particles Combining, 102
Pauli Exclusion Principle, 130
Penergy
  definition of, 3
  density, 7
  density scale, 10
  expansion of, 10
  NIB-state, 6
  origin of, 4
Penergy density
  chart of, 30
  in relation to SMBH, 20
  related to Relativity, 159
Penergy medium
  a frenzy of activity, 54
  as Quantum Potential, 60
  the definition of, 42
Penrose, Roger, 5
Perlov, Delia, 142
photoelectric effect, 172
photon
  absorbed and emitted, 135
  configuration, 132
  mimicking electric field, 134
  mimicking magnetic field, 134
  travels in straight line, 157
photon detection device, 58
photon energy
  definition of, 198
photon evolution, 141
photon polarization
  entanglement, 182
pilot wave theory, 59, 174
Pion, 102, 128, 137
plasma state, 118
potential energy
  definition of, 197
precession
  in electron, 119
Preons, 91
pure energy, 3, *See* Penergy
  interspatial medium, 42
Pure-Energy Blackhole
  evidence for, 21

Q

Quantum Electrodynamics, 65
Quantum Field Theory
  ubiquitous fields, 48

INDEX

Quantum Mechanics, 166, 173, 178, 186
   the definition of, 51
quantum potential, 60
Quantum Theory, 171, 176
   wave-particle duality, 56
quark
   charge value, 122
   configuration, 125
   decay, 88, 125
   evolution, 128
   formation of charge values, 123

R

radiation energy
   definition of, 196
Randall, Lisa, 41, 191
reality
   of Quantum Theory, 186
   the origin of, 170
reductionist philosophy, 194
Rees, Martin, 15
relativity, 159
renormalization, 65
Rezzolla, Luciano, 13, 150, 215
Rubin, Vera, 145

S

Schrodinger equation, 185
Schumm, Bruce, 38, 91, 104, 105
Schwarzschild, Karl, 5, 13, 149
Seife, Charles, 26
Siegel, Ethan, 84
Silk, Joseph, 112
singularity
   inside blackhole, 13
Singularity Problem, 5
Smith, Timothy Paul, 83, 111

Smolin, Lee, 170, 177, 190
space & time
   why relative, 155
spacetime, 32
spin
   and all forces, 82
   importance of, 34
spin signature, 38
spin-space, 105
spiral galaxies
   mergers of, 15
Standard Model, 106, 114
   and the weak force, 87
stellar collapse, 12
Still, Ben, 4, 107, 110
String Theory, 191
strong force
   per Standard Model, 83
   per Tor Model, 85
Sub-A's
   definition of, 2
   Level 3 particles, 103
supermassive blackhole
   alternative creation theory, 17
   theory of, 14
Supersymmetry, 191
Sutter, Paul, 23, 43

T

theory of everything, 208
thermodynamics, 11
time dilation, 160
   bouncing light scenario, 158
Tor Chains, 71
Tor Model, 215
   definition of, 32
   hidden variables, 184
   particle location, 187
   quarks made from Tryks, 122
   wave function, 186

INDEX

Tor spin
    and gravitational field, 80
    and gyroscopic effect, 109
    and magnetic force, 69
    and strong force, 85
    spin dynamics, 85
Tor-Chain
    bundle, 73
Torons, 32
Tors
    created in band of Penergy, 104
    Level 1 particles, 102
    not subject to annihilation, 75
Tryk-Chain wobble, 119
Tryks
    illustration of, 76
    Level 2 particles, 102
Tucker, Wallace H., 25
Turner, Ben, 163
Tyson, Neil deGrasse, 4

U

universe
    charge neutral, 36
    initial conditions, 1

V

vacuum of space, 67, 87
    energy borrowed, 41
Vilenkin, Alex, 142
virtual particles
    existence of, 41
    in Tor Model, 67
virtual photons, 64

W

W and Z bosons, 87
wave particle duality
    Pilot Wave Theory, 59
wavefunction, 173, 185
weak force, 87
Whiteson, Daniel, 42, 113, 115
Wilczek, Frank, 109, 129, 171
wobble
    in Muon magnetic moment, 119
work energy, 3, 196

About the Author: Robert J. Conover

After earning a degree in philosophy, Bob took up studying particle physics and cosmology with a passion. Intrigued by the new discoveries taking place in particle physics and their ramifications on our changing picture of the early universe, he was drawn into finding answers to the many questions, conflicts, and unknowns that arose from those discoveries.

A New Vision of the Early Universe is a culmination of forty years of research and study of the writings of many particle physicists, cosmologists, and mathematicians. The depth of Bob's research, his attention to detail, and his philosophical approach to problem solving made him the ideal person to write this book. Unrelentless in his pursuit of a viable vision of the early universe, Bob has amassed additional material supporting a broader perspective on how the universe evolved, necessitating this Second Edition.

Bob has retired from a career as a civil investigator and mediation negotiator, during which time he authored *Strategic Tort-Mediation Negotiation*. Bob resides with his wife Ellen in San Luis Obispo, California, where he continues to research, write, create board games, and cultivate the minds of their eight grandchildren.

Made in the USA
Middletown, DE
04 April 2024